Jordan Canonical Form: Application to Differential Equations

Jordan Canonical Form: Application to Differential Equations
Steven H. Weintraub

ISBN: 978-3-031-01267-9   paperback
ISBN: 978-3-031-02395-8   ebook

DOI 10.1007/978-3-031-02395-8

A Publication in the Springer series
SYNTHESIS LECTURES ON MATHEMATICS AND STATISTICS

Lecture #2
Series Editor: Steven G. Krantz, Washington University, St. Louis

Series ISSN
Synthesis Lectures on Mathematics and Statistics
ISSN pending.

Jordan Canonical Form:
Application to
Differential Equations

Steven H. Weintraub
Lehigh University

SYNTHESIS LECTURES ON MATHEMATICS AND STATISTICS #2

ABSTRACT

Jordan Canonical Form (JCF) is one of the most important, and useful, concepts in linear algebra. In this book we develop JCF and show how to apply it to solving systems of differential equations. We first develop JCF, including the concepts involved in it–eigenvalues, eigenvectors, and chains of generalized eigenvectors. We begin with the diagonalizable case and then proceed to the general case, but we do not present a complete proof. Indeed, our interest here is not in JCF per se, but in one of its important applications. We devote the bulk of our attention in this book to showing how to apply JCF to solve systems of constant-coefficient first order differential equations, where it is a very effective tool. We cover all situations–homogeneous and inhomogeneous systems; real and complex eigenvalues. We also treat the closely related topic of the matrix exponential. Our discussion is mostly confined to the 2-by-2 and 3-by-3 cases, and we present a wealth of examples that illustrate all the possibilities in these cases (and of course, a wealth of exercises for the reader).

KEYWORDS

Jordan Canonical Form, linear algebra, differential equations, eigenvalues, eigenvectors, generalized eigenvectors, matrix exponential

Contents

Preface

Jordan Canonical Form (JCF) is one of the most important, and useful, concepts in linear algebra. In this book, we develop JCF and show how to apply it to solving systems of differential equations.

In Chapter 1, we develop JCF. We do not prove the existence of JCF in general, but we present the ideas that go into it—eigenvalues and (chains of generalized) eigenvectors. In Section 1.1, we treat the diagonalizable case, and in Section 1.2, we treat the general case. We develop all possibilities for 2-by-2 and 3-by-3 matrices, and illustrate these by examples.

In Chapter 2, we apply JCF. We show how to use JCF to solve systems $Y' = AY + G(x)$ of constant-coefficient first-order linear differential equations. In Section 2.1, we consider homogeneous systems $Y' = AY$. In Section 2.2, we consider homogeneous systems when the characteristic polynomial of A has complex roots (in which case an additional step is necessary). In Section 2.3, we consider inhomogeneous systems $Y' = AY + G(x)$ with $G(x)$ nonzero. In Section 2.4, we develop the matrix exponential e^{Ax} and relate it to solutions of these systems. Also in this chapter we provide examples that illustrate all the possibilities in the 2-by-2 and 3-by-3 cases.

Appendix A has background material. Section A.1 gives background on coordinates for vectors and matrices for linear transformations. Section A.2 derives the basic properties of the complex exponential function. This material is relegated to the Appendix so that readers who are unfamiliar with these notions, or who are willing to take them on faith, can skip it and still understand the material in Chapters 1 and 2.

Our numbering system for results is fairly standard: Theorem 2.1, for example, is the first Theorem found in Section 2 of Chapter 1.

As is customary in textbooks, we provide the answers to the odd-numbered exercises here. *Instructors* may contact me at shw2@lehigh.edu and I will supply the answers to all of the exercises.

Steven H. Weintraub
Lehigh University
Bethlehem, PA USA
July 2008

CHAPTER 1

Jordan Canonical Form

1.1 THE DIAGONALIZABLE CASE

Although, for simplicity, most of our examples will be over the real numbers (and indeed over the rational numbers), we will consider that *all of our vectors and matrices are defined over the complex numbers* $\mathbb{C}$. It is only with this assumption that the theory of Jordan Canonical Form (JCF) works completely. See Remark 1.9 for the key reason why.

Definition 1.1. If $v \neq 0$ is a vector such that, for some λ,

$$Av = \lambda v \,,$$

then v is an *eigenvector* of A associated to the *eigenvalue* λ.

Example 1.2. Let A be the matrix $A = \begin{bmatrix} 5 & -7 \\ 2 & -4 \end{bmatrix}$. Then, as you can check, if $v_1 = \begin{bmatrix} 7 \\ 2 \end{bmatrix}$, then $Av_1 = 3v_1$, so v_1 is an eigenvector of A with associated eigenvalue 3, and if $v_2 = \begin{bmatrix} 1 \\ 1 \end{bmatrix}$, then $Av_2 = -2v_2$, so v_2 is an eigenvector of A with associated eigenvalue -2.

We note that the definition of an eigenvalue/eigenvector can be expressed in an alternate form. Here I denotes the identity matrix:

$$Av = \lambda v$$
$$Av = \lambda I v$$
$$(A - \lambda I)v = 0 \,.$$

For an eigenvalue λ of A, we let E_λ denote the *eigenspace* of λ,

$$E_\lambda = \{v \mid Av = \lambda v\} = \{v \mid (A - \lambda I)v = 0\} = \mathrm{Ker}(A - \lambda I) \,.$$

(The kernel $\mathrm{Ker}(A - \lambda I)$ is also known as the nullspace $\mathrm{NS}(A - \lambda I)$.)

We also note that this alternate formulation helps us find eigenvalues and eigenvectors. For if $(A - \lambda I)v = 0$ for a nonzero vector v, the matrix $A - \lambda I$ must be singular, and hence its determinant must be 0. This leads us to the following definition.

Definition 1.3. The *characteristic polynomial* of a matrix A is the polynomial $\det(\lambda I - A)$.

Remark 1.4.   This is the customary definition of the characteristic polynomial. But note that, if A is an n-by-n matrix, then the matrix $\lambda I - A$ is obtained from the matrix $A - \lambda I$ by multiplying each of its n rows by -1, and hence $\det(\lambda I - A) = (-1)^n \det(A - \lambda I)$. In practice, it is most convenient to work with $A - \lambda I$ in finding eigenvectors—this minimizes arithmetic—and when we come to find chains of generalized eigenvectors in Section 1.2, it is (almost) essential to use $A - \lambda I$, as using $\lambda I - A$ would introduce lots of spurious minus signs.

Example 1.5.   Returning to the matrix $A = \begin{bmatrix} 5 & -7 \\ 2 & -4 \end{bmatrix}$ of Example 1.2, we compute that $\det(\lambda I - A) = \lambda^2 - \lambda - 6 = (\lambda - 3)(\lambda + 2)$, so A has eigenvalues 3 and -2. Computation then shows that the eigenspace $E_3 = \text{Ker}(A - 3I)$ has basis $\left\{ \begin{bmatrix} 7 \\ 2 \end{bmatrix} \right\}$, and that the eigenspace $E_{-2} = \text{Ker}(A - (-2)I)$ has basis $\left\{ \begin{bmatrix} 1 \\ 1 \end{bmatrix} \right\}$.

We now introduce two important quantities associated to an eigenvalue of a matrix A.

Definition 1.6.   Let a be an eigenvalue of a matrix A. The *algebraic multiplicity* of the eigenvalue a is alg-mult(a) = the multiplicity of a as a root of the characteristic polynomial $\det(\lambda I - A)$. The *geometric multiplicity* of the eigenvalue a is geom-mult(a) = the dimension of the eigenspace E_a.

It is common practice to use the word *multiplicity* (without a qualifier) to mean algebraic multiplicity.

We have the following relationship between these two multiplicities.

Lemma 1.7.   *Let a be an eigenvalue of a matrix A. Then*

$$1 \leq \text{geom-mult}(a) \leq \text{alg-mult}(a) .$$

Proof. By the definition of an eigenvalue, there is at least one eigenvector v with eigenvalue a, and so E_a contains the nonzero vector v, and hence $\dim(E_a) \geq 1$.

For the proof that geom-mult(a) $\leq$ alg-mult(a), see Lemma 1.12 in Appendix A.   □

Corollary 1.8.   *Let a be an eigenvalue of A and suppose that a has algebraic multiplicity* 1. *Then a also has geometric multiplicity* 1.

Proof. In this case, applying Lemma 1.7, we have

$$1 \leq \text{geom-mult}(a) \leq \text{alg-mult}(a) = 1 ,$$

so geom-mult(a) = 1.   □

Remark 1.9. Let A be an n-by-n matrix. Then its characteristic polynomial $\det(\lambda I - A)$ has degree n. *Since we are considering A to be defined over the complex numbers*, we may apply the Fundamental Theorem of Algebra, which states that an nth degree polynomial has n roots, counting multiplicities. Hence, we see that, for any n-by-n matrix A, the sum of the algebraic multiplicities of the eigenvalues of A is equal to n.

Lemma 1.10. *Let A be an n-by-n matrix. The following are equivalent:*
(1) For each eigenvalue a of A, geom-mult(a) = alg-mult(a).
(2) The sum of the geometric multiplicities of the eigenvalues of A is equal to n.

Proof. Let A have eigenvalues $a_1, a_2, \ldots, a_m$. For each i between 1 and m, let $s_i = $ geom-mult(a_i) and $t_i = $ alg-mult(a_i). Then, by Lemma 1.7, $s_i \leq t_i$ for each i, and by Remark 1.9, $\sum_{i=1}^m t_i = n$. Thus, if $s_i = t_i$ for each i, then $\sum_{i=1}^m s_i = n$, while if $s_i < t_i$ for some i, then $\sum_{i=1}^m s_i < n$. □

Proposition 1.11. *(1) Let $a_1, a_2, \ldots, a_m$ be distinct eigenvalues of A (i.e., $a_i \neq a_j$ for $i \neq j$). For each i between 1 and m, let v_i be an associated eigenvector. Then $\{v_1, v_2, \ldots, v_m\}$ is a linearly independent set of vectors.*
(2) More generally, let $a_1, a_2, \ldots, a_m$ be distinct eigenvalues of A. For each i between 1 and m, let S_i be a linearly independent set of eigenvectors associated to a_i. Then $S = S_1 \cup \ldots S_m$ is a linearly independent set of vectors.

Proof. (1) Suppose we have a linear combination $0 = c_1 v_1 + c_2 v_2 + \ldots + c_m v_m$. We need to show that $c_i = 0$ for each i. To do this, we begin with an observation: If v is an eigenvector of A associated to the eigenvalue a, and b is any scalar, then $(A - bI)v = Av - bv = av - bv = (a - b)v$. (Note that this answer is 0 if $a = b$ and nonzero if $a \neq b$.)

We now go to work, multiplying our original relation by $(A - a_m I)$. Of course, $(A - a_m I)0 = 0$, so:

$$0 = (A - a_m I)(c_1 v_1 + c_2 v_2 + \ldots + c_{m-2} v_{m-2} + c_{m-1} v_{m-1} + c_m v_m)$$
$$= c_1 (A - a_m I)v_1 + c_2 (A - a_m I)v_2 + \ldots$$
$$+ c_{m-2}(A - a_m I)v_{m-2} + c_{m-1}(A - a_m I)v_{m-1} + c_m (A - a_m I)v_m$$
$$= c_1 (a_1 - a_m)v_1 + c_2 (a_2 - a_m)v_2 + \ldots$$
$$+ c_{m-2}(a_{m-2} - a_m)v_{m-2} + c_{m-1}(a_{m-1} - a_m)v_{m-1} .$$

We now multiply this relation by $(A - a_{m-1}I)$. Again, $(A - a_{m-1}I)0 = 0$, so:

$$0 = (A - a_{m-1}I)(c_1(a_1 - a_m)v_1 + c_2(a_2 - a_m)v_2 + \ldots$$
$$+ c_{m-2}(a_{m-2} - a_m)v_{m-2} + c_{m-1}(a_{m-1} - a_m)v_{m-1})$$
$$= c_1(a_1 - a_m)(A - a_{m-1}I)v_1 + c_2(a_2 - a_m)(A - a_{m-1}I)v_2 + \ldots$$
$$+ c_{m-2}(a_{m-2} - a_m)(A - a_{m-1}I)v_{m-2} + c_{m-1}(a_{m-1} - a_m)(A - a_{m-1}I)v_{m-1}$$
$$= c_1(a_1 - a_m)(a_1 - a_{m-1})v_1 + c_2(a_2 - a_m)(a_2 - a_{m-1})v_2 + \ldots$$
$$+ c_{m-2}(a_{m-2} - a_m)(a_{m-2} - a_{m-1})v_{m-2} \, .$$

Proceed in this way, until at the last step we multiply by $(A - a_2I)$. We then obtain:

$$0 = c_1(a_1 - a_2) \cdots (a_1 - a_{m-1})(a_1 - a_m)v_1 \, .$$

But $v_1 \neq 0$, as by definition an eigenvector is nonzero. Also, the product $(a_1 - a_2) \cdots (a_1 - a_{m-1})(a_1 - a_m)$ is a product of nonzero numbers and is hence nonzero. Thus, we must have $c_1 = 0$.

Proceeding in the same way, multiplying our original relation by $(A - a_mI)$, $(A - a_{m-1}I)$, $(A - a_3I)$, and finally by $(A - a_1I)$, we obtain $c_2 = 0$, and, proceeding in this vein, we obtain $c_i = 0$ for all i, and so the set $\{v_1, v_2, \ldots, v_m\}$ is linearly independent.

(2) To avoid complicated notation, we will simply prove this when $m = 2$ (which illustrates the general case). Thus, let $m = 2$, let $S_1 = \{v_{1,1}, \ldots, v_{1,i_1}\}$ be a linearly independent set of eigenvectors associated to the eigenvalue a_1 of A, and let $S_2 = \{v_{2,1}, \ldots, v_{2,i_2}\}$ be a linearly independent set of eigenvectors associated to the eigenvalue a_2 of A. Then $S = \{v_{1,1}, \ldots, v_{1,i_1}, v_{2,1}, \ldots, v_{2,i_2}\}$. We want to show that S is a linearly independent set. Suppose we have a linear combination $0 = c_{1,1}v_{1,1} + \ldots + c_{1,i_1}v_{1,i_1} + c_{2,1}v_{2,1} + \ldots + c_{2,i_2}v_{2,i_2}$. Then:

$$0 = c_{1,1}v_{1,1} + \ldots + c_{1,i_1}v_{1,i_1} + c_{2,1}v_{2,1} + \ldots + c_{2,i_2}v_{2,i_2}$$
$$= (c_{1,1}v_{1,1} + \ldots + c_{1,i_1}v_{1,i_1}) + (c_{2,1}v_{2,1} + \ldots + c_{2,i_2}v_{2,i_2})$$
$$= v_1 + v_2$$

where $v_1 = c_{1,1}v_{1,1} + \ldots + c_{1,i_1}v_{1,i_1}$ and $v_2 = c_{2,1}v_{2,1} + \ldots + c_{2,i_2}v_{2,i_2}$. But v_1 is a vector in E_{a_1}, so $Av_1 = a_1v_1$; similarly, v_2 is a vector in E_{a_2}, so $Av_2 = a_2v_2$. Then, as in the proof of part (1),

$$0 = (A - a_2I)0 = (A - a_2I)(v_1 + v_2) = (A - a_2I)v_1 + (A - a_2I)v_2$$
$$= (a_1 - a_2)v_1 + 0 = (a_1 - a_2)v_1$$

so $0 = v_1$; similarly, $0 = v_2$. But $0 = v_1 = c_{1,1}v_{1,1} + \ldots + c_{1,i_1}v_{1,i_1}$ implies $c_{1,1} = \ldots c_{1,i_1} = 0$, as, by hypothesis, $\{v_{1,1}, \ldots, v_{1,i_1}\}$ is a linearly independent set; similarly, $0 = v_2$ implies $c_{2,1} = \ldots = c_{2,i_2} = 0$. Thus, $c_{1,1} = \ldots = c_{1,i_1} = c_{2,1} = \ldots = c_{2,i_2} = 0$ and S is linearly independent, as claimed. $\square$

Definition 1.12. Two square matrices A and B are *similar* if there is an invertible matrix P with $A = PBP^{-1}$.

Definition 1.13. A square matrix A is *diagonalizable* if A is similar to a diagonal matrix.

Here is the main result of this section.

Theorem 1.14. *Let A be an n–by–n matrix over the complex numbers. Then A is diagonalizable if and only if, for each eigenvalue a of A, geom-mult(a) = alg-mult(a). In that case, $A = PJP^{-1}$ where J is a diagonal matrix whose entries are the eigenvalues of A, each appearing according to its algebraic multiplicity, and P is a matrix whose columns are eigenvectors forming bases for the associated eigenspaces.*

Proof. We give a proof by direct computation here. For a more conceptual proof, see Theorem 1.10 in Appendix A.

First let us suppose that for each eigenvalue a of A, geom-mult(a) = alg-mult(a).

Let A have eigenvalues $a_1, a_2, \ldots, a_n$. Here we do not insist that the a_i's are distinct; rather, each eigenvalue appears the same number of times as its algebraic multiplicity. Then J is the diagonal matrix

$$J = \left[\, j_1 \,\middle|\, j_2 \,\middle|\, \cdots \,\middle|\, j_n \,\right]$$

and we see that j_i, the ith column of J, is the vector

$$j_i = \begin{bmatrix} 0 \\ \vdots \\ 0 \\ a_i \\ 0 \\ \vdots \end{bmatrix},$$

with a_i in the ith position, and 0 elsewhere.

We have

$$P = \left[\, v_1 \,\middle|\, v_2 \,\middle|\, \cdots \,\middle|\, v_n \,\right],$$

a matrix whose columns are eigenvectors forming bases for the associated eigenspaces. By hypothesis, geom-mult(a) = alg-mult(a) for each eigenvector a of A, so there are as many columns of P that are eigenvectors for the eigenvalue a as there are diagonal entries of J that are equal to a. Furthermore, by Lemma 1.10, the matrix P indeed has n columns.

We first show by direct computation that $AP = PJ$. Now

$$AP = A\left[\, v_1 \,\middle|\, v_2 \,\middle|\, \cdots \,\middle|\, v_n \,\right]$$

so the ith column of AP is Av_i. But

$$Av_i = a_i v_i$$

as v_i is an eigenvector of A with associated eigenvalue a_i.

On the other hand,

$$PJ = \left[v_1 \middle| v_2 \middle| \dots \middle| v_n \right] J$$

and the ith column of PJ is Pj_i,

$$Pj_i = \left[v_1 \middle| v_2 \middle| \dots \middle| v_n \right] j_i .$$

Remembering what the vector j_i is, and multiplying, we see that

$$Pj_i = a_i v_i$$

as well.

Thus, every column of AP is equal to the corresponding column of PJ, so

$$AP = PJ .$$

By Proposition 1.11, the columns of the square matrix P are linearly independent, so P is invertible. Multiplying on the right by P^{-1}, we see that

$$A = PJP^{-1} ,$$

completing the proof of this half of the Theorem.

Now let us suppose that A is diagonalizable, $A = PJP^{-1}$. Then $AP = PJ$. We use the same notation for P and J as in the first half of the proof. Then, as in the first half of the proof, we compute AP and PJ column-by-column, and we see that the ith column of AP is Av_i and that the ith column of PJ is $a_i v_i$, for each i. Hence, $Av_i = a_i v_i$ for each i, and so v_i is an eigenvector of A with associated eigenvalue a_i.

For each eigenvalue a of A, there are as many columns of P that are eigenvectors for a as there are diagonal entries of J that are equal to a, and these vectors form a basis for the eigenspace associated of the eigenvalue a, so we see that for each eigenvalue a of A, geom-mult(a) = alg-mult(a), completing the proof. □

For a general matrix A, the condition in Theorem 1.14 may or may not be satisfied, i.e., some but not all matrices are diagonalizable. But there is one important case when this condition is automatic.

Corollary 1.15. *Let A be an n–by–n matrix over the complex numbers all of whose eigenvalues are distinct (i.e., whose characteristic polynomial has no repeated roots). Then A is diagonalizable.*

Proof. By hypothesis, for each eigenvalue a of A, alg-mult$(a) = 1$. But then, by Corollary 1.8, for each eigenvalue a of A, geom-mult$(a) =$ alg-mult(a), so the hypothesis of Theorem 1.14 is satisfied.

$\square$

Example 1.16.   Let A be the matrix $A = \begin{bmatrix} 5 & -7 \\ 2 & -4 \end{bmatrix}$ of Examples 1.2 and 1.5. Then, referring to Example 1.5, we see

$$\begin{bmatrix} 5 & -7 \\ 2 & -4 \end{bmatrix} = \begin{bmatrix} 7 & 1 \\ 2 & 1 \end{bmatrix} \begin{bmatrix} 3 & 0 \\ 0 & -2 \end{bmatrix} \begin{bmatrix} 7 & 1 \\ 2 & 1 \end{bmatrix}^{-1} .$$

As we have indicated, we have developed this theory over the complex numbers, as JFC works best over them. But there is an analog of our results over the real numbers—we just have to require that all the eigenvalues of A are real. Here is the basic result on diagonalizability.

Theorem 1.17.   *Let A be an n-by-n real matrix. Then A is diagonalizable if and only if all the eigenvalues of A are real numbers, and, for each eigenvalue a of A, geom-mult$(a) =$ alg-mult(a). In that case, $A = PJP^{-1}$ where J is a diagonal matrix whose entries are the eigenvalues of A, each appearing according to its algebraic multiplicity (and hence is a real matrix), and P is a real matrix whose columns are eigenvectors forming bases for the associated eigenspaces.*

1.2   THE GENERAL CASE

Let us begin this section by describing what a matrix in JCF looks like. A matrix in JCF is composed of "Jordan blocks," so we first see what a single Jordan block looks like.

Definition 2.1.   A k-by-k *Jordan block* associated to the eigenvalue λ is a k-by-k matrix of the form

$$J = \begin{bmatrix} \lambda & 1 & & & & \\ & \lambda & 1 & & & \\ & & \lambda & 1 & & \\ & & & \ddots & \ddots & \\ & & & & \lambda & 1 \\ & & & & & \lambda \end{bmatrix} .$$

In other words, a Jordan block is a matrix with all the diagonal entries equal to each other, all the entries immediately above the diagonal equal to 1, and all the other entries equal to 0.

Definition 2.2. A matrix J in *Jordan Canonical Form* (JCF) is a block diagonal matrix

$$
J = \begin{bmatrix} J_1 & & & & \\ & J_2 & & & \\ & & J_3 & & \\ & & & \ddots & \\ & & & & J_\ell \end{bmatrix}
$$

with each J_i a Jordan block.

Remark 2.3. Note that every diagonal matrix is a matrix in JCF, with each Jordan block a 1-by-1 block.

In order to understand and be able to use JCF, we must introduce a new concept, that of a generalized eigenvector.

Definition 2.4. If $v \neq 0$ is a vector such that, for some λ,

$$
(A - \lambda I)^k(v) = 0
$$

for some positive integer k, then v is a *generalized eigenvector* of A associated to the eigenvalue λ. The smallest k with $(A - \lambda I)^k(v) = 0$ is the *index* of the generalized eigenvector v.

Let us note that if v is a generalized eigenvector of index 1, then

$$
\begin{aligned}
(A - \lambda I)(v) &= 0 \\
(A)v &= (\lambda I)v \\
Av &= \lambda v
\end{aligned}
$$

and so v is an (ordinary) eigenvector.

Recall that, for an eigenvalue λ of A, E_λ denotes the eigenspace of λ,

$$
E_\lambda = \{v \mid Av = \lambda v\} = \{v \mid (A - \lambda I)v = 0\} .
$$

We let $\tilde{E}_\lambda$ denote the *generalized eigenspace* of λ,

$$
\tilde{E}_\lambda = \{v \mid (A - \lambda I)^k(v) = 0 \text{ for some } k\} .
$$

It is easy to check that $\tilde{E}_\lambda$ is a subspace.

Since every eigenvector is a generalized eigenvector, we see that

$$E_\lambda \subseteq \tilde{E}_\lambda \ .$$

The following result (which we shall not prove) is an important fact about generalized eigenspaces.

Proposition 2.5. *Let λ be an eigenvalue of the n-by-n matrix A of algebraic multiplicity m. Then, $\tilde{E}_\lambda$ is a subspace of $\mathbb{C}^n$ of dimension m.*

Example 2.6. Let A be the matrix $A = \begin{bmatrix} 0 & 1 \\ -4 & 4 \end{bmatrix}$. Then, as you can check, if $u = \begin{bmatrix} 1 \\ 2 \end{bmatrix}$, then $(A - 2I)u = 0$, so u is an eigenvector of A with associated eigenvalue 2 (and hence a generalized eigenvector of index 1 of A with associated eigenvalue 2). On the other hand, if $v = \begin{bmatrix} 1 \\ 0 \end{bmatrix}$, then $(A - 2I)^2 v = 0$ but $(A - 2I)v \neq 0$, so v is a generalized eigenvector of index 2 of A with associated eigenvalue 2.

In this case, as you can check, the vector u is a basis for the eigenspace E_2, so $E_2 = \{ cu \mid c \in \mathbb{C} \}$ is one dimensional.

On the other hand, u and v are both generalized eigenvectors associated to the eigenvalue 2, and are linearly independent (the equation $c_1 u + c_2 v = 0$ only has the solution $c_1 = c_2 = 0$, as you can readily check), so $\tilde{E}_2$ has dimension at least 2. Since $\tilde{E}_2$ is a subspace of $\mathbb{C}^2$, it must have dimension exactly 2, and $\tilde{E}_2 = \mathbb{C}^2$ (and $\{u, v\}$ is indeed a basis for $\mathbb{C}^2$).

Let us next consider a generalized eigenvector v_k of index k associated to an eigenvalue λ, and set

$$v_{k-1} = (A - \lambda I)v_k \ .$$

We claim that v_{k-1} is a generalized eigenvector of index $k - 1$ associated to the eigenvalue λ. To see this, note that

$$(A - \lambda I)^{k-1} v_{k-1} = (A - \lambda I)^{k-1}(A - \lambda I)v_k = (A - \lambda I)^k v_k = 0$$

but

$$(A - \lambda I)^{k-2} v_{k-1} = (A - \lambda I)^{k-2}(A - \lambda I)v_k = (A - \lambda I)^{k-1} v_k \neq 0 \ .$$

Proceeding in this way, we may set

$$v_{k-2} = (A - \lambda I)v_{k-1} = (A - \lambda I)^2 v_k$$
$$v_{k-3} = (A - \lambda I)v_{k-2} = (A - \lambda I)^2 v_{k-1} = (A - \lambda I)^3 v_k$$
$$\vdots$$
$$v_1 = (A - \lambda I)v_2 = \cdots = (A - \lambda I)^{k-1} v_k$$

and note that each v_i is a generalized eigenvector of index i associated to the eigenvalue λ. A collection of generalized eigenvectors obtained in this way gets a special name:

Definition 2.7. If $\{v_1, \ldots, v_k\}$ is a set of generalized eigenvectors associated to the eigenvalue λ of A, such that v_k is a generalized eigenvector of index k and also

$$v_{k-1} = (A - \lambda I)v_k, \quad v_{k-2} = (A - \lambda I)v_{k-1}, \quad v_{k-3} = (A - \lambda I)v_{k-2},$$
$$\cdots, \quad v_2 = (A - \lambda I)v_3, \quad v_1 = (A - \lambda I)v_2,$$

then $\{v_1, \ldots, v_k\}$ is called a *chain* of generalized eigenvectors of length k. The vector v_k is called the *top* of the chain and the vector v_1 (which is an ordinary eigenvector) is called the *bottom* of the chain.

Example 2.8. Let us return to Example 2.6. We saw there that $v = \begin{bmatrix} 1 \\ 0 \end{bmatrix}$ is a generalized eigenvector of index 2 of $A = \begin{bmatrix} 0 & 1 \\ -4 & 4 \end{bmatrix}$ associated to the eigenvalue 2. Let us set $v_2 = v = \begin{bmatrix} 1 \\ 0 \end{bmatrix}$. Then $v_1 = (A - 2I)v_2 = \begin{bmatrix} -2 \\ -4 \end{bmatrix}$ is a generalized eigenvector of index 1 (i.e., an ordinary eigenvector), and $\{v_1, v_2\}$ is a chain of length 2.

Remark 2.9. It is important to note that a chain of generalized eigenvectors $\{v_1, \ldots, v_k\}$ is entirely determined by the vector v_k at the top of the chain. For once we have chosen v_k, there are no other choices to be made: the vector v_{k-1} is determined by the equation $v_{k-1} = (A - \lambda I)v_k$; then the vector v_{k-2} is determined by the equation $v_{k-2} = (A - \lambda I)v_{k-1}$, etc.

With this concept in hand, let us return to JCF. As we have seen, a matrix J in JCF has a number of blocks $J_1, J_2, \ldots, J_\ell$, called Jordan blocks, along the diagonal. Let us begin our analysis with the case when J consists of a single Jordan block. So suppose J is a k-by-k matrix

$$J = \begin{bmatrix} \lambda & 1 & & & & \\ & \lambda & 1 & & 0 & \\ & & \lambda & 1 & & \\ & & & \ddots & \ddots & \\ & 0 & & & \lambda & 1 \\ & & & & & \lambda \end{bmatrix}.$$

Then,

$$J - \lambda I = \begin{bmatrix} 0 & 1 & & & & \\ & 0 & 1 & & & \\ & & 0 & 1 & & \\ & & & \ddots & \ddots & \\ & & & & 0 & 1 \\ & & & & & 0 \end{bmatrix}.$$

Let $e_1 = \begin{bmatrix} 1 \\ 0 \\ 0 \\ \vdots \\ 0 \end{bmatrix}$, $e_2 = \begin{bmatrix} 0 \\ 1 \\ 0 \\ \vdots \\ 0 \end{bmatrix}$, $e_3 = \begin{bmatrix} 0 \\ 0 \\ 1 \\ \vdots \\ 0 \end{bmatrix}$, ..., $e_k = \begin{bmatrix} 0 \\ 0 \\ 0 \\ \vdots \\ 1 \end{bmatrix}$.

Then direct calculation shows:

$$(J - \lambda I)e_k = e_{k-1}$$
$$(J - \lambda I)e_{k-1} = e_{k-2}$$
$$\vdots$$
$$(J - \lambda I)e_2 = e_1$$
$$(J - \lambda I)e_1 = 0$$

and so we see that $\{e_1, \ldots, e_k\}$ is a chain of generalized eigenvectors. We also note that $\{e_1, \ldots, e_k\}$ is a basis for $\mathbb{C}^k$, and so

$$\tilde{E}_\lambda = \mathbb{C}^k.$$

We first see that the situation is very analogous when we consider any k-by-k matrix with a single chain of generalized eigenvectors of length k.

Proposition 2.10. *Let $\{v_1, \ldots, v_k\}$ be a chain of generalized eigenvectors of length k associated to the eigenvalue λ of a matrix A. Then $\{v_1, \ldots, v_k\}$ is linearly independent.*

Proof. Suppose we have a linear combination

$$c_1 v_1 + c_2 v_2 + \cdots + c_{k-1} v_{k-1} + c_k v_k = 0.$$

We must show each $c_i = 0$.

By the definition of a chain, $v_{k-i} = (A - \lambda I)^i v_k$ for each i, so we may write this equation as

$$c_1 (A - \lambda I)^{k-1} v_k + c_2 (A - \lambda I)^{k-2} v_k + \cdots + c_{k-1}(A - \lambda I)v_k + c_k v_k = 0.$$

Now let us multiply this equation on the left by $(A - \lambda I)^{k-1}$. Then we obtain the equation

$$c_1(A - \lambda I)^{2k-2}v_k + c_2(A - \lambda I)^{2k-3}v_k + \cdots + c_{k-1}(A - \lambda I)^k v_k + c_k(A - \lambda I)^{k-1}v_k = 0 .$$

Now $(A - \lambda I)^{k-1}v_k = v_1 \neq 0$. However, $(A - \lambda I)^k v_k = 0$, and then also $(A - \lambda I)^{k+1}v_k = (A - \lambda I)(A - \lambda I)^k v_k = (A - \lambda I)(0) = 0$, and then similarly $(A - \lambda I)^{k+2}v_k = 0, \ldots , (A - \lambda I)^{2k-2}v_k = 0$, so every term except the last one is zero and this equation becomes

$$c_k v_1 = 0 .$$

Since $v_1 \neq 0$, this shows $c_k = 0$, so our linear combination is

$$c_1 v_1 + c_2 v_2 + \cdots + c_{k-1}v_{k-1} = 0 .$$

Repeat the same argument, this time multiplying by $(A - \lambda I)^{k-2}$ instead of $(A - \lambda I)^{k-1}$. Then we obtain the equation

$$c_{k-1}v_1 = 0 ,$$

and, since $v_1 \neq 0$, this shows that $c_{k-1} = 0$ as well. Keep going to get

$$c_1 = c_2 = \cdots = c_{k-1} = c_k = 0 ,$$

so $\{v_1, \ldots , v_k\}$ is linearly independent.   □

Theorem 2.11. *Let A be a k-by-k matrix and suppose that $\mathbb{C}^k$ has a basis $\{v_1, \ldots , v_k\}$ consisting of a single chain of generalized eigenvectors of length k associated to an eigenvalue a. Then*

$$A = PJP^{-1}$$

where

$$J = \begin{bmatrix} a & 1 & & & & \\ & a & 1 & & & \\ & & a & 1 & & \\ & & & \ddots & \ddots & \\ & & & & a & 1 \\ & & & & & a \end{bmatrix}$$

is a matrix consisting of a single Jordan block and

$$P = \begin{bmatrix} v_1 & v_2 & \cdots & v_k \end{bmatrix}$$

is a matrix whose columns are generalized eigenvectors forming a chain.

Proof. We give a proof by direct computation here. (Note the similarity of this proof to the proof of Theorem 1.14.) For a more conceptual proof, see Theorem 1.11 in Appendix A.

Let P be the given matrix. We will first show by direct computation that $AP = PJ$. It will be convenient to write

$$J = \left[\; j_1 \;\middle|\; j_2 \;\middle|\; \cdots \;\middle|\; j_k \;\right]$$

and we see that j_i, the ith column of J, is the vector

$$j_i = \begin{bmatrix} 0 \\ \vdots \\ 1 \\ a \\ 0 \\ \vdots \end{bmatrix}$$

with 1 in the $(i-1)^{st}$ position, a in the ith position, and 0 elsewhere.

We show that $AP = PJ$ by showing that their corresponding columns are equal. Now

$$AP = A\left[\; v_1 \;\middle|\; v_2 \;\middle|\; \cdots \;\middle|\; v_k \;\right]$$

so the ith column of AP is Av_i. But

$$\begin{aligned} Av_i &= (A - aI + aI)v_i \\ &= (A - aI)v_i + aIv_i \\ &= v_{i-1} + av_i \text{ for } i > 1, \; = av_i \text{ for } i = 1 \;. \end{aligned}$$

On the other hand,

$$PJ = \left[\; v_1 \;\middle|\; v_2 \;\middle|\; \cdots \;\middle|\; v_k \;\right]J$$

and the ith column of PJ is Pj_i,

$$Pj_i = \left[\; v_1 \;\middle|\; v_2 \;\middle|\; \cdots \;\middle|\; v_k \;\right]j_i \;.$$

Remembering what the vector j_i is, and multiplying, we see that

$$Pj_i = v_{i-1} + av_i \text{ for } i > 1, \; = av_i \text{ for } i = 1$$

as well.

Thus, every column of AP is equal to the corresponding column of PJ, so

$$AP = PJ.$$

But Proposition 2.10 shows that the columns of P are linearly independent, so P is invertible. Multiplying on the right by P^{-1}, we see that

$$A = PJP^{-1}.$$

□

Example 2.12. Applying Theorem 2.11 to the matrix $A = \begin{bmatrix} 0 & 1 \\ -4 & 4 \end{bmatrix}$ of Examples 2.6 and 2.8, we see that

$$\begin{bmatrix} 0 & 1 \\ -4 & 4 \end{bmatrix} = \begin{bmatrix} -2 & 1 \\ -4 & 0 \end{bmatrix} \begin{bmatrix} 2 & 1 \\ 0 & 2 \end{bmatrix} \begin{bmatrix} -2 & 1 \\ -4 & 0 \end{bmatrix}^{-1}.$$

Here is the key theorem to which we have been heading. This theorem is one of the most important (and useful) theorems in linear algebra.

Theorem 2.13. *Let A be any square matrix defined over the complex numbers. Then A is similar to a matrix in Jordan Canonical Form. More precisely, $A = PJP^{-1}$, for some matrix J in Jordan Canonical Form. The diagonal entries of J consist of eigenvalues of A, and P is an invertible matrix whose columns are chains of generalized eigenvectors of A.*

Proof. (Rough outline) In general, the JCF of a matrix A does not consist of a single block, but will have a number of blocks, of varying sizes and associated to varying eigenvalues.

But in this situation we merely have to "assemble" the various blocks (to get the matrix J) and the various chains of generalized eigenvectors (to get a basis and hence the matrix P). Actually, the word "merely" is a bit misleading, as the argument that we can do so is, in fact, a subtle one, and we shall not give it here. □

In lieu of proving Theorem 2.13, we shall give a number of examples that illustrate the situation. In fact, in order to avoid complicated notation we shall merely illustrate the situation for 2-by-2 and 3-by-3 matrices.

Theorem 2.14. *Let A be a 2-by-2 matrix. Then one of the following situations applies:*

(i) *A has two eigenvalues, a and b, each of algebraic multiplicity* 1. *Let u be an eigenvector associated to the eigenvalue a and let v be an eigenvector associated to the eigenvalue b. Then* $A = PJP^{-1}$ *with*

$$J = \begin{bmatrix} a & 0 \\ 0 & b \end{bmatrix} \quad and \quad P = \begin{bmatrix} u & v \end{bmatrix}.$$

(Note, in this case, A is diagonalizable.)

(ii) *A has a single eigenvalue a of algebraic multiplicity* 2.

(a) *A has two linearly independent eigenvectors u and v.*
    Then $A = PJP^{-1}$ *with*

$$J = \begin{bmatrix} a & 0 \\ 0 & a \end{bmatrix} \quad and \quad P = \begin{bmatrix} u & v \end{bmatrix}.$$

(Note, in this case, A is diagonalizable. In fact, in this case $E_a = \mathbb{C}^2$ and A itself is the matrix $\begin{bmatrix} a & 0 \\ 0 & a \end{bmatrix}$.)

(b) *A has a single chain $\{v_1, v_2\}$ of generalized eigenvectors. Then* $A = PJP^{-1}$ *with*

$$J = \begin{bmatrix} a & 1 \\ 0 & a \end{bmatrix} \quad and \quad P = \begin{bmatrix} v_1 & v_2 \end{bmatrix}.$$

Theorem 2.15.   *Let A be a 3-by-3 matrix. Then one of the following situations applies:*

(i) *A has three eigenvalues, a, b, and c, each of algebraic multiplicity* 1. *Let u be an eigenvector associated to the eigenvalue a, v be an eigenvector associated to the eigenvalue b, and w be an eigenvector associated to the eigenvalue c. Then* $A = PJP^{-1}$ *with*

$$J = \begin{bmatrix} a & 0 & 0 \\ 0 & b & 0 \\ 0 & 0 & c \end{bmatrix} \quad and \quad P = \begin{bmatrix} u & v & w \end{bmatrix}.$$

(Note, in this case, A is diagonalizable.)

(ii) *A has an eigenvalue a of algebraic multiplicity* 2 *and an eigenvalue b of algebraic multiplicity* 1.

(a) *A has two independent eigenvectors, u and v, associated to the eigenvalue a. Let w be an eigenvector associated to the eigenvalue b. Then* $A = PJP^{-1}$ *with*

$$J = \begin{bmatrix} a & 0 & 0 \\ 0 & a & 0 \\ 0 & 0 & b \end{bmatrix} \quad and \quad P = \begin{bmatrix} u & v & w \end{bmatrix}.$$

(Note, in this case, A is diagonalizable.)

(b) A has a single chain $\{u_1, u_2\}$ of generalized eigenvectors associated to the eigenvalue a. Let v be an eigenvector associated to the eigenvalue b. Then $A = PJP^{-1}$ with

$$J = \begin{bmatrix} a & 1 & 0 \\ 0 & a & 0 \\ 0 & 0 & b \end{bmatrix} \quad and \quad P = \left[u_1 \,\middle|\, u_2 \,\middle|\, v \right].$$

(iii) A has a single eigenvalue a of algebraic multiplicity 3.

(a) A has three linearly independent eigenvectors, u, v, and w. Then $A = PJP^{-1}$ with

$$J = \begin{bmatrix} a & 0 & 0 \\ 0 & a & 0 \\ 0 & 0 & a \end{bmatrix} \quad and \quad P = \left[u \,\middle|\, v \,\middle|\, w \right].$$

(Note, in this case, A is diagonalizable. In fact, in this case $E_a = \mathbb{C}^3$ and A itself is the matrix $\begin{bmatrix} a & 0 & 0 \\ 0 & a & 0 \\ 0 & 0 & a \end{bmatrix}$.)

(b) A has a chain $\{u_1, u_2\}$ of generalized eigenvectors and an eigenvector v with $\{u_1, u_2, v\}$ linearly independent. Then $A = PJP^{-1}$ with

$$J = \begin{bmatrix} a & 1 & 0 \\ 0 & a & 0 \\ 0 & 0 & a \end{bmatrix} \quad and \quad P = \left[u_1 \,\middle|\, u_2 \,\middle|\, v \right].$$

(c) A has a single chain $\{u_1, u_2, u_3\}$ of generalized eigenvectors. Then $A = PJP^{-1}$ with

$$J = \begin{bmatrix} a & 1 & 0 \\ 0 & a & 1 \\ 0 & 0 & a \end{bmatrix} \quad and \quad P = \left[u_1 \,\middle|\, u_2 \,\middle|\, u_3 \right].$$

Remark 2.16.   Suppose that A has JCF $J = aI$, a scalar multiple of the identity matrix. Then $A = PJP^{-1} = P(aI)P^{-1} = a(PIP^{-1}) = aI = J$. This justifies the parenthetical remark in Theorems 2.14 (ii) (a) and 2.15 (iii) (a).

Remark 2.17.   Note that Theorems 2.14 (i), 2.14 (ii) (a), 2.15 (i), 2.15 (ii) (a), and 2.15 (iii) (a) are all special cases of Theorem 1.14, and in fact Theorems 2.14 (i) and 2.15 (i) are both special cases of Corollary 1.15.

Now we would like to apply Theorems 2.14 and 2.15. In order to do so, we need to have an effective method to determine which of the cases we are in, and we give that here (without proof).

Definition 2.18. Let λ be an eigenvalue of A. Then for any positive integer i,

$$E_\lambda^i = \{v \mid (A - \lambda I)^i(v) = 0\}$$
$$= \mathrm{Ker}((A - \lambda I)^i) .$$

Note that E_λ^i consists of generalized eigenvectors of index at most i (and the 0 vector), and is a subspace. Note also that

$$E_\lambda = E_\lambda^1 \subseteq E_\lambda^2 \subseteq \ldots \subseteq \tilde{E}_\lambda .$$

In general, the JCF of A is determined by the dimensions of all the spaces E_λ^i, but this determination can be a bit complicated. For eigenvalues of multiplicity at most 3, however, the situation is simpler—we need only consider the eigenspaces E_λ. This is a consequence of the following general result.

Proposition 2.19. *Let λ be an eigenvalue of A. Then the number of blocks in the JCF of A corresponding to λ is equal to $\dim E_\lambda$, i.e., to the geometric multiplicity of λ.*

Proof. (Outline) Suppose there are ℓ such blocks. Since each block corresponds to a chain of generalized eigenvectors, there are ℓ such chains. Now the bottom of the chain is an (ordinary) eigenvector, so we get ℓ eigenvectors in this way. It can be shown that these ℓ eigenvectors are always linearly independent and that they always span E_λ, i.e., that they are a basis of E_λ. Thus, E_λ has a basis consisting of ℓ vectors, so $\dim E_\lambda = \ell$. ☐

We can now determine the JCF of 1-by-1, 2-by-2, and 3-by-3 matrices, using the following consequences of this proposition.

Corollary 2.20. *Let λ be an eigenvalue of A of algebraic multiplicity 1. Then $\dim E_\lambda^1 = 1$, i.e., a has geometric multiplicity 1, and the submatrix of the JCF of A corresponding to the eigenvalue λ is a single 1-by-1 block.*

Corollary 2.21. *Let λ be an eigenvalue of A of algebraic multiplicity 2. Then there are the following possibilities:*

(a) $\dim E_\lambda^1 = 2$, i.e., a has geometric multiplicity 2. In this case, the submatrix of the JCF of A corresponding to the eigenvalue λ consists of two 1-by-1 blocks.

(b) $\dim E_\lambda^1 = 1$, *i.e., a has geometric multiplicity* 1. *Also,* $\dim E_\lambda^2 = 2$. *In this case, the submatrix of the JCF of A corresponding to the eigenvalue λ consists of a single 2-by-2 block.*

Corollary 2.22.   *Let λ be an eigenvalue of A of algebraic multiplicity* 3. *Then there are the following possibilities:*

(a) $\dim E_\lambda^1 = 3$, *i.e., a has geometric multiplicity* 3. *In this case, the submatrix of the JCF of A corresponding to the eigenvalue λ consists of three 1-by-1 blocks.*

(b) $\dim E_\lambda^1 = 2$, *i.e., a has geometric multiplicity* 2. *Also,* $\dim E_\lambda^2 = 3$. *In this case, the submatrix of the Jordan Canonical Form of A corresponding to the eigenvalue λ consists of a 2-by-2 block and a 1-by-1 block.*

(c) $\dim E_\lambda^1 = 1$, *i.e., a has geometric multiplicity* 1. *Also,* $\dim E_\lambda^2 = 2$, *and* $\dim E_\lambda^3 = 3$. *In this case, the submatrix of the Jordan Canonical Form of A corresponding to the eigenvalue λ consists of a single 3-by-3 block.*

Now we shall do several examples.

Example 2.23. $A = \begin{bmatrix} 2 & -3 & -3 \\ 2 & -2 & -2 \\ -2 & 1 & 1 \end{bmatrix}$.

A has characteristic polynomial $\det(\lambda I - A) = (\lambda + 1)(\lambda)(\lambda - 2)$. Thus, A has eigenvalues $-1, 0,$ and 2, each of multiplicity one, and so we are in the situation of Theorem 2.15 (i). Computation shows that the eigenspace $E_{-1} = \text{Ker}(A - (-I))$ has basis $\left\{ \begin{bmatrix} 1 \\ 0 \\ 1 \end{bmatrix} \right\}$, the eigenspace $E_0 = \text{Ker}(A)$ has basis $\left\{ \begin{bmatrix} 0 \\ 1 \\ -1 \end{bmatrix} \right\}$, and the eigenspace $E_2 = \text{Ker}(A - 2I)$ has basis $\left\{ \begin{bmatrix} -1 \\ -1 \\ 1 \end{bmatrix} \right\}$. Hence, we see that

$$\begin{bmatrix} 2 & -3 & -3 \\ 2 & -2 & -2 \\ -2 & 1 & 1 \end{bmatrix} = \begin{bmatrix} 1 & 0 & -1 \\ 0 & -1 & -1 \\ 1 & 1 & 1 \end{bmatrix} \begin{bmatrix} -1 & 0 & 0 \\ 0 & 0 & 0 \\ 0 & 0 & 2 \end{bmatrix} \begin{bmatrix} 1 & 0 & -1 \\ 0 & -1 & -1 \\ 1 & 1 & 1 \end{bmatrix}^{-1} .$$

Example 2.24. $A = \begin{bmatrix} 3 & 1 & 1 \\ 2 & 4 & 2 \\ 1 & 1 & 3 \end{bmatrix}$.

A has characteristic polynomial $\det(\lambda I - A) = (\lambda - 2)^2(\lambda - 6)$. Thus, A has an eigenvalue 2 of multiplicity 2 and an eigenvalue 6 of multiplicity 1. Computation shows that the eigenspace

$$E_2 = \text{Ker}(A - 2I) \text{ has basis } \left\{ \begin{bmatrix} -1 \\ 1 \\ 0 \end{bmatrix}, \begin{bmatrix} -1 \\ 0 \\ 1 \end{bmatrix} \right\}, \text{ so dim } E_2 = 2 \text{ and we are in the situation of}$$

Corollary 2.21 (a). Further computation shows that the eigenspace $E_6 = \text{Ker}(A - 6I)$ has basis $\left\{ \begin{bmatrix} 1 \\ 2 \\ 1 \end{bmatrix} \right\}$. Hence, we see that

$$\begin{bmatrix} 3 & 1 & 1 \\ 2 & 4 & 2 \\ 1 & 1 & 3 \end{bmatrix} = \begin{bmatrix} -1 & -1 & 1 \\ 1 & 0 & 2 \\ 0 & 1 & 1 \end{bmatrix} \begin{bmatrix} 2 & 0 & 0 \\ 0 & 2 & 0 \\ 0 & 0 & 6 \end{bmatrix} \begin{bmatrix} -1 & -1 & 1 \\ 1 & 0 & 2 \\ 0 & 1 & 1 \end{bmatrix}^{-1}.$$

Example 2.25. $A = \begin{bmatrix} 2 & 1 & 1 \\ 2 & 1 & -2 \\ -1 & 0 & -2 \end{bmatrix}.$

A has characteristic polynomial $\det(\lambda I - A) = (\lambda + 1)^2(\lambda - 3)$. Thus, A has an eigenvalue -1 of multiplicity 2 and an eigenvalue 3 of multiplicity 1. Computation shows that the eigenspace

$$E_{-1} = \text{Ker}(A - (-I)) \text{ has basis } \left\{ \begin{bmatrix} -1 \\ 2 \\ 1 \end{bmatrix} \right\} \text{ so dim } E_{-1} = 1 \text{ and we are in the situation of Corol-}$$

lary 2.21 (b). Then we further compute that $E_{-1}^2 = \text{Ker}((A - (-I))^2)$ has basis $\left\{ \begin{bmatrix} -1 \\ 2 \\ 0 \end{bmatrix}, \begin{bmatrix} 0 \\ 0 \\ 1 \end{bmatrix} \right\}$,

therefore is two-dimensional, as we expect. More to the point, we may choose any generalized eigen-vector of index 2, i.e., any vector in E_{-1}^2 that is not in E_{-1}^1, as the top of a chain. We choose $u_2 = \begin{bmatrix} 0 \\ 0 \\ 1 \end{bmatrix}$, and then we have $u_1 = (A - (-I))u_2 = \begin{bmatrix} 1 \\ -2 \\ -1 \end{bmatrix}$, and $\{u_1, u_2\}$ form a chain.

We also compute that, for the eigenvalue 3, the eigenspace E_3 has basis $\left\{ v = \begin{bmatrix} -5 \\ -6 \\ 1 \end{bmatrix} \right\}$.

Hence, we see that

$$\begin{bmatrix} 2 & 1 & 1 \\ 2 & 1 & -2 \\ -1 & 0 & 2 \end{bmatrix} = \begin{bmatrix} 1 & 0 & -5 \\ -2 & 0 & -6 \\ -1 & 1 & 1 \end{bmatrix} \begin{bmatrix} -1 & 1 & 0 \\ 0 & -1 & 0 \\ 0 & 0 & 3 \end{bmatrix} \begin{bmatrix} 1 & 0 & -5 \\ -2 & 0 & -6 \\ -1 & 1 & 1 \end{bmatrix}^{-1}.$$

Example 2.26. $A = \begin{bmatrix} 2 & 1 & 1 \\ -2 & -1 & -2 \\ 1 & 1 & 2 \end{bmatrix}$.

A has characteristic polynomial $\det(\lambda I - A) = (\lambda - 1)^3$, so A has one eigenvalue 1 of multiplicity three. Computation shows that $E_1 = \text{Ker}(A - I)$ has basis $\left\{ \begin{bmatrix} -1 \\ 0 \\ 1 \end{bmatrix}, \begin{bmatrix} -1 \\ 1 \\ 0 \end{bmatrix} \right\}$, so $\dim E_1 = 2$ and we are in the situation of Corollary 2.22 (b). Computation then shows that $\dim E_1^2 = 3$ (i.e., $(A - I)^2 = 0$ and E_1^2 is all of $\mathbb{C}^3$) with basis $\left\{ \begin{bmatrix} 1 \\ 0 \\ 0 \end{bmatrix}, \begin{bmatrix} 0 \\ 1 \\ 0 \end{bmatrix}, \begin{bmatrix} 0 \\ 0 \\ 1 \end{bmatrix} \right\}$. We may choose u_2 to be any vector in E_1^2 that is not in E_1^1, and we shall choose $u_2 = \begin{bmatrix} 1 \\ 0 \\ 0 \end{bmatrix}$. Then $u_1 = (A - I)u_2 = \begin{bmatrix} 1 \\ -2 \\ 1 \end{bmatrix}$, and $\{u_1, u_2\}$ form a chain. For the third vector, v, we may choose any vector in E_1 such that $\{u_1, v\}$ is linearly independent. We choose $v = \begin{bmatrix} -1 \\ 0 \\ 1 \end{bmatrix}$. Hence, we see that

$$\begin{bmatrix} 2 & 1 & 1 \\ -2 & -1 & 2 \\ 1 & 1 & 2 \end{bmatrix} = \begin{bmatrix} 1 & 1 & -1 \\ -2 & 0 & 0 \\ 1 & 0 & 1 \end{bmatrix} \begin{bmatrix} 1 & 1 & 0 \\ 0 & 1 & 0 \\ 0 & 0 & 1 \end{bmatrix} \begin{bmatrix} 1 & 1 & -1 \\ -2 & 0 & 0 \\ 1 & 0 & 1 \end{bmatrix}^{-1}.$$

Example 2.27. $A = \begin{bmatrix} 5 & 0 & 1 \\ 1 & 1 & 0 \\ -7 & 1 & 0 \end{bmatrix}$.

A has characteristic polynomial $\det(\lambda I - A) = (\lambda - 2)^3$, so A has one eigenvalue 2 of multiplicity three. Computation shows that $E_2 = \text{Ker}(A - 2I)$ has basis $\left\{ \begin{bmatrix} -1 \\ -1 \\ 3 \end{bmatrix} \right\}$, so $\dim E_2^1 = 1$ and we are in the situation of Corollary 2.22 (c). Then computation shows that $E_2^2 = \text{Ker}((A - 2I)^2)$ has basis $\left\{ \begin{bmatrix} -1 \\ 0 \\ 2 \end{bmatrix}, \begin{bmatrix} -1 \\ 2 \\ 0 \end{bmatrix} \right\}$. (Note that $\begin{bmatrix} -1 \\ -1 \\ 3 \end{bmatrix} = 3/2 \begin{bmatrix} -1 \\ 0 \\ 2 \end{bmatrix} + 1/2 \begin{bmatrix} -1 \\ 2 \\ 0 \end{bmatrix}$.) Computation then shows that dim $E_2^3 = 3$ (i.e., $(A - 2I)^3 = 0$ and E_2^3 is all of $\mathbb{C}^3$) with basis $\left\{ \begin{bmatrix} 1 \\ 0 \\ 0 \end{bmatrix}, \begin{bmatrix} 0 \\ 1 \\ 0 \end{bmatrix}, \begin{bmatrix} 0 \\ 0 \\ 1 \end{bmatrix} \right\}$.

We may choose u_3 to be any vector in $\mathbb{C}^3$ that is not in E_2^2, and we shall choose $u_3 = \begin{bmatrix} 1 \\ 0 \\ 0 \end{bmatrix}$. Then

$$u_2 = (A - 2I)u_3 = \begin{bmatrix} 3 \\ 1 \\ -7 \end{bmatrix} \text{ and } u_1 = (A - 2I)u_2 = \begin{bmatrix} 2 \\ 2 \\ -6 \end{bmatrix}, \text{ and then } \{u_1, u_2, u_3\} \text{ form a chain.}$$

Hence, we see that

$$\begin{bmatrix} 5 & 0 & 1 \\ 1 & 1 & 0 \\ -7 & 1 & 0 \end{bmatrix} = \begin{bmatrix} 2 & 3 & 1 \\ 2 & 1 & 0 \\ -6 & -7 & 0 \end{bmatrix} \begin{bmatrix} 2 & 1 & 0 \\ 0 & 2 & 1 \\ 0 & 0 & 2 \end{bmatrix} \begin{bmatrix} 2 & 3 & 1 \\ 2 & 1 & 0 \\ -6 & -7 & 0 \end{bmatrix}^{-1}.$$

As we have mentioned, we need to work over the complex numbers in order for the theory of JCF to fully apply. But there is an analog over the real numbers, and we conclude this section by stating it.

Theorem 2.28.   *Let A be a real square matrix (i.e., a square matrix with all entries real numbers), and suppose that all of the eigenvalues of A are real numbers. Then A is similar to a real matrix in Jordan Canonical Form. More precisely, $A = PJP^{-1}$ with P and J real matrices, for some matrix J in Jordan Canonical Form. The diagonal entries of J consist of eigenvalues of A, and P is an invertible matrix whose columns are chains of generalized eigenvectors of A.*

EXERCISES FOR CHAPTER 1

For each matrix A, write $A = PJP^{-1}$ with P an invertible matrix and J a matrix in JCF.

1. $A = \begin{bmatrix} 75 & 56 \\ -90 & -67 \end{bmatrix}$, $\det(\lambda I - A) = (\lambda - 3)(\lambda - 5)$.

2. $A = \begin{bmatrix} -50 & 99 \\ -20 & 39 \end{bmatrix}$, $\det(\lambda I - A) = (\lambda + 6)(\lambda + 5)$.

3. $A = \begin{bmatrix} -18 & 9 \\ -49 & 24 \end{bmatrix}$, $\det(\lambda I - A) = (\lambda - 3)^2$.

4. $A = \begin{bmatrix} 1 & 1 \\ -16 & 9 \end{bmatrix}$, $\det(\lambda I - A) = (\lambda - 5)^2$.

5. $A = \begin{bmatrix} 2 & 1 \\ -25 & 12 \end{bmatrix}$, $\det(\lambda I - A) = (\lambda - 7)^2$.

6. $A = \begin{bmatrix} -15 & 9 \\ -25 & 15 \end{bmatrix}$, $\det(\lambda I - A) = \lambda^2$.

7. $A = \begin{bmatrix} 1 & 0 & 0 \\ 1 & 2 & -3 \\ 1 & -1 & 0 \end{bmatrix}$, $\det(\lambda I - A) = (\lambda + 1)(\lambda - 1)(\lambda - 3)$.

8. $A = \begin{bmatrix} 3 & 0 & 2 \\ 1 & 3 & 1 \\ 0 & 1 & 1 \end{bmatrix}$, $\det(\lambda I - A) = (\lambda - 1)(\lambda - 2)(\lambda - 4)$.

9. $A = \begin{bmatrix} 5 & 8 & 16 \\ 4 & 1 & 8 \\ -4 & -4 & -11 \end{bmatrix}$, $\det(\lambda I - A) = (\lambda + 3)^2(\lambda - 1)$.

10. $A = \begin{bmatrix} 4 & 2 & 3 \\ -1 & 1 & -3 \\ 2 & 4 & 9 \end{bmatrix}$, $\det(\lambda I - A) = (\lambda - 3)^2(\lambda - 8)$.

11. $A = \begin{bmatrix} 5 & 2 & 1 \\ -1 & 2 & -1 \\ -1 & -2 & 3 \end{bmatrix}$, $\det(\lambda I - A) = (\lambda - 4)^2(\lambda - 2)$.

12. $A = \begin{bmatrix} 8 & -3 & -3 \\ 4 & 0 & -2 \\ -2 & 1 & 3 \end{bmatrix}$, $\det(\lambda I - A) = (\lambda - 2)^2(\lambda - 7)$.

13. $A = \begin{bmatrix} -3 & 1 & -1 \\ -7 & 5 & -1 \\ -6 & 6 & -2 \end{bmatrix}$, $\det(\lambda I - A) = (\lambda + 2)^2(\lambda - 4)$.

14. $A = \begin{bmatrix} 3 & 0 & 0 \\ 9 & -5 & -18 \\ -4 & 4 & 12 \end{bmatrix}$, $\det(\lambda I - A) = (\lambda - 3)^2(\lambda - 4)$.

15. $A = \begin{bmatrix} -6 & 9 & 0 \\ -6 & 6 & -2 \\ 9 & -9 & 3 \end{bmatrix}$, $\det(\lambda I - A) = \lambda^2(\lambda - 3)$.

16. $A = \begin{bmatrix} -18 & 42 & 168 \\ 1 & -7 & -40 \\ -2 & 6 & 27 \end{bmatrix}$, $\det(\lambda I - A) = (\lambda - 3)^2(\lambda + 4)$.

17. $A = \begin{bmatrix} -1 & 1 & -1 \\ -10 & 6 & -5 \\ -6 & 3 & -2 \end{bmatrix}$, $\det(\lambda I - A) = (\lambda - 1)^3$.

18. $A = \begin{bmatrix} 0 & -4 & 1 \\ 2 & -6 & 1 \\ 4 & -8 & 0 \end{bmatrix}$, $\det(\lambda I - A) = (\lambda + 2)^3$.

19. $A = \begin{bmatrix} -4 & 1 & 2 \\ -5 & 1 & 3 \\ -7 & 2 & 3 \end{bmatrix}$, $\det(\lambda I - A) = \lambda^3$.

20. $A = \begin{bmatrix} -4 & -2 & 5 \\ -1 & -1 & 1 \\ -2 & -1 & 2 \end{bmatrix}$, $\det(\lambda I - A) = (\lambda + 1)^3$.

CHAPTER 2

Solving Systems of Linear Differential Equations

2.1 HOMOGENEOUS SYSTEMS WITH CONSTANT COEFFICIENTS

We will now see how to use Jordan Canonical Form (JCF) to solve systems $Y' = AY$. We begin by describing the strategy we will follow throughout this section.

Consider the matrix system

$$Y' = AY .$$

Step 1. Write $A = PJP^{-1}$ with J in JCF, so the system becomes

$$Y' = (PJP^{-1})Y$$
$$Y' = PJ(P^{-1}Y)$$
$$P^{-1}Y' = J(P^{-1}Y)$$
$$(P^{-1}Y)' = J(P^{-1}Y) .$$

(Note that, since P^{-1} is a constant matrix, we have that $(P^{-1}Y)' = P^{-1}Y'$.)

Step 2. Set $Z = P^{-1}Y$, so this system becomes

$$Z' = JZ$$

and solve this system for Z.

Step 3. Since $Z = P^{-1}Y$, we have that

$$Y = PZ$$

is the solution to our original system.

Examining this strategy, we see that we already know how to carry out Step 1, and also that Step 3 is very easy—it is just matrix multiplication. Thus, the key to success here is being able to carry out Step 2. This is where JCF comes in. As we shall see, it is (relatively) easy to solve $Z' = JZ$ when J is a matrix in JCF.

You will note that throughout this section, in solving $Z' = JZ$, we write the solution as $Z = M_Z C$, where M_Z is a matrix of functions, called the *fundamental matrix* of the system, and C is a vector of arbitrary constants. The reason for this will become clear later. (See Remarks 1.12 and 1.14.)

Although it is not logically necessary—we may regard a diagonal matrix as a matrix in JCF in which all the Jordan blocks are 1-by-1 blocks—it is illuminating to handle the case when J is diagonal first. Here the solution is very easy.

Theorem 1.1. *Let J be a k-by-k diagonal matrix,*

$$
J = \begin{bmatrix}
a_1 & & & & & & \\
& a_2 & & & & 0 & \\
& & a_3 & & & & \\
& & & \ddots & & & \\
& 0 & & & a_{k-1} & \\
& & & & & & a_k
\end{bmatrix}.
$$

Then the system $Z' = JZ$ has the solution

$$
Z = \begin{bmatrix}
e^{a_1 x} & & & & & & \\
& e^{a_2 x} & & & 0 & \\
& & e^{a_3 x} & & & \\
& & & \ddots & & \\
& 0 & & & e^{a_{k-1} x} & \\
& & & & & e^{a_k x}
\end{bmatrix} \quad C = M_Z C
$$

where $C = \begin{bmatrix} c_1 \\ c_2 \\ \vdots \\ c_k \end{bmatrix}$ *is a vector of arbitrary constants $c_1, c_2, \ldots, c_k$.*

Proof. Multiplying out, we see that the system $Z' = JZ$ is just the system

$$
\begin{bmatrix} z_1' \\ z_2' \\ \vdots \\ z_k' \end{bmatrix} = \begin{bmatrix} a_1 z_1 \\ a_2 z_2 \\ \vdots \\ a_k z_k \end{bmatrix}.
$$

But this system is "uncoupled", i.e., the equation for z_i' only involves z_i and none of the other functions. Now this equation is very familiar. In general, the differential equation $z' = az$ has solution $z = ce^{ax}$, and applying that here we find that $Z' = JZ$ has solution

$$Z = \begin{bmatrix} c_1 e^{a_1 x} \\ c_2 e^{a_2 x} \\ \vdots \\ c_k e^{a_k x} \end{bmatrix},$$

which is exactly the above product $M_Z C$. $\qquad\square$

Example 1.2. Consider the system

$$Y' = AY \quad \text{where} \quad A = \begin{bmatrix} 5 & -7 \\ 2 & -4 \end{bmatrix}.$$

We saw in Example 1.16 in Chapter 1 that $A = PJP^{-1}$ with

$$P = \begin{bmatrix} 7 & 1 \\ 2 & 1 \end{bmatrix} \quad \text{and} \quad J = \begin{bmatrix} 3 & 0 \\ 0 & -2 \end{bmatrix}.$$

Then $Z' = JZ$ has solution

$$Z = \begin{bmatrix} e^{3x} & 0 \\ 0 & e^{-2x} \end{bmatrix} \begin{bmatrix} c_1 \\ c_2 \end{bmatrix} = M_Z C = \begin{bmatrix} c_1 e^{3x} \\ c_2 e^{-2x} \end{bmatrix}$$

and so $Y = PZ = PM_Z C$, i.e.,

$$Y = \begin{bmatrix} 7 & 1 \\ 2 & 1 \end{bmatrix} \begin{bmatrix} e^{3x} & 0 \\ 0 & e^{-2x} \end{bmatrix} \begin{bmatrix} c_1 \\ c_2 \end{bmatrix}$$

$$= \begin{bmatrix} 7e^{3x} & e^{-2x} \\ 2e^{3x} & e^{-2x} \end{bmatrix} \begin{bmatrix} c_1 \\ c_2 \end{bmatrix}$$

$$= \begin{bmatrix} 7c_1 e^{3x} + c_2 e^{-2x} \\ 2c_1 e^{3x} + c_2 e^{-2x} \end{bmatrix}.$$

Example 1.3. Consider the system

$$Y' = AY \quad \text{where} \quad A = \begin{bmatrix} 2 & -3 & -3 \\ 2 & -2 & -2 \\ -2 & 1 & 1 \end{bmatrix}.$$

We saw in Example 2.23 in Chapter 1 that $A = PJP^{-1}$ with

$$P = \begin{bmatrix} 1 & 0 & -1 \\ 0 & -1 & -1 \\ 1 & 1 & 1 \end{bmatrix} \text{ and } J = \begin{bmatrix} -1 & 0 & 0 \\ 0 & 0 & 0 \\ 0 & 0 & 2 \end{bmatrix}.$$

Then $Z' = JZ$ has solution

$$Z = \begin{bmatrix} e^{-x} & 0 & 0 \\ 0 & 1 & 0 \\ 0 & 0 & e^{2x} \end{bmatrix} \begin{bmatrix} c_1 \\ c_2 \\ c_3 \end{bmatrix} = M_Z C$$

and so $Y = PZ = PM_Z C$, i.e.,

$$Y = \begin{bmatrix} 1 & 0 & -1 \\ 0 & -1 & -1 \\ 1 & 1 & 1 \end{bmatrix} \begin{bmatrix} e^{-x} & 0 & 0 \\ 0 & 1 & 0 \\ 0 & 0 & e^{2x} \end{bmatrix} \begin{bmatrix} c_1 \\ c_2 \\ c_3 \end{bmatrix}$$

$$= \begin{bmatrix} e^{-x} & 0 & -e^{2x} \\ 0 & -1 & -e^{2x} \\ e^{-x} & 1 & e^{2x} \end{bmatrix} \begin{bmatrix} c_1 \\ c_2 \\ c_3 \end{bmatrix}$$

$$= \begin{bmatrix} c_1 e^{-x} & - c_3 e^{2x} \\ -c_2 - c_3 e^{2x} \\ c_1 e^{-x} + c_2 + c_3 e^{2x} \end{bmatrix}.$$

We now see how to use JCF to solve systems $Y' = AY$ where the coefficient matrix A is not diagonalizable.

The key to understanding systems is to investigate a system $Z' = JZ$ where J is a matrix consisting of a single Jordan block. Here the solution is not as easy as in Theorem 1.1, but it is still not too hard.

Theorem 1.4. *Let J be a k-by-k Jordan block with eigenvalue a,*

$$J = \begin{bmatrix} a & 1 & & & & \\ & a & 1 & & 0 & \\ & & a & 1 & & \\ & & & \ddots & \ddots & \\ & 0 & & & a & 1 \\ & & & & & a \end{bmatrix}.$$

Then the system $Z' = JZ$ has the solution

$$Z = e^{ax} \begin{bmatrix} 1 & x & x^2/2! & x^3/3! & \cdots & x^{k-1}/(k-1)! \\ & 1 & x & x^2/2! & \cdots & x^{k-2}/(k-2)! \\ & & 1 & x & \cdots & x^{k-3}/(k-3)! \\ & & & \ddots & & \vdots \\ & & & & & x \\ & & & & & 1 \end{bmatrix} C = M_Z C$$

where $C = \begin{bmatrix} c_1 \\ c_2 \\ \vdots \\ c_k \end{bmatrix}$ is a vector of arbitrary constants $c_1, c_2, \ldots, c_k$.

Proof. We will prove this in the cases $k = 1, 2$, and 3, which illustrate the pattern. As you will see, the proof is a simple application of the standard technique for solving first-order linear differential equations.

The case $k = 1$:   Here we are considering the system

$$[z_1'] = [a][z_1]$$

which is nothing other than the differential equation

$$z_1' = az_1 \, .$$

This differential equation has solution

$$z_1 = c_1 e^{ax} \, ,$$

which we can certainly write as

$$[z_1] = e^{az}[1][c_1] \, .$$

The case $k = 2$:   Here we are considering the system

$$\begin{bmatrix} z_1' \\ z_2' \end{bmatrix} = \begin{bmatrix} a & 1 \\ 0 & a \end{bmatrix} \begin{bmatrix} z_1 \\ z_2 \end{bmatrix} \, ,$$

which is nothing other than the pair of differential equations

$$\begin{aligned} z_1' &= az_1 + z_2 \\ z_2' &= \quad az_2 \, . \end{aligned}$$

We recognize the second equation as having the solution

$$z_2 = c_2 e^{ax}$$

and we substitute this into the first equation to get

$$z_1' = az_1 + c_2 e^{ax} .$$

To solve this, we rewrite this as

$$z_1' - az_1 = c_2 e^{ax}$$

and recognize that this differential equation has integrating factor e^{-ax}. Multiplying by this factor, we find

$$e^{-ax}(z_1' - az_1) = c_2$$
$$(e^{-ax}z_1)' = c_2$$
$$e^{-ax}z_1 = \int c_2 \, dx = c_1 + c_2 x$$

so

$$z_1 = e^{ax}(c_1 + c_2 x) .$$

Thus, our solution is

$$z_1 = e^{ax}(c_1 + c_2 x)$$
$$z_2 = \phantom{e^{ax}(c_1 + } e^{ax} c_2 ,$$

which we see we can rewrite as

$$\begin{bmatrix} z_1 \\ z_2 \end{bmatrix} = e^{ax} \begin{bmatrix} 1 & x \\ 0 & 1 \end{bmatrix} \begin{bmatrix} c_1 \\ c_2 \end{bmatrix} .$$

The case $k = 3$:   Here we are considering the system

$$\begin{bmatrix} z_1' \\ z_2' \\ z_3' \end{bmatrix} = \begin{bmatrix} a & 1 & 0 \\ 0 & a & 1 \\ 0 & 0 & a \end{bmatrix} \begin{bmatrix} z_1 \\ z_2 \\ z_3 \end{bmatrix} ,$$

which is nothing other than the triple of differential equations

$$z_1' = az_1 + z_2$$
$$z_2' = az_2 + z_3$$
$$z_3' = az_3 .$$

If we just concentrate on the last two equations, we see we are in the $k = 2$ case. Referring to that case, we see that our solution is

$$z_2 = e^{ax}(c_2 + c_3 x)$$
$$z_3 = \phantom{e^{ax}(} e^{ax} c_3 \, .$$

Substituting the value of z_2 into the equation for z_1, we obtain

$$z_1' = az_1 + e^{ax}(c_2 + c_3 x) \, .$$

To solve this, we rewrite this as

$$z_1' - az_1 = e^{ax}(c_2 + c_3 x)$$

and recognize that this differential equation has integrating factor e^{-ax}. Multiplying by this factor, we find

$$e^{-ax}(z_1' - az_1) = c_2 + c_3 x$$
$$(e^{-ax} z_1)' = c_2 + c_3 x$$
$$e^{-ax} z_1 = \int (c_2 + c_3 x) \, dx = c_1 + c_2 x + c_3 (x^2/2)$$

so

$$z_1 = e^{ax}(c_1 + c_2 x + c_3(x^2/2)) \, .$$

Thus, our solution is

$$z_1 = e^{ax}(c_1 + c_2 x + c_3(x^2/2))$$
$$z_2 = \phantom{e^{ax}(} e^{ax}(c_2 + c_3 x)$$
$$z_3 = \phantom{e^{ax}(c_2} e^{ax} c_3 \, ,$$

which we see we can rewrite as

$$\begin{bmatrix} z_1 \\ z_2 \\ z_3 \end{bmatrix} = e^{ax} \begin{bmatrix} 1 & x & x^2/2 \\ 0 & 1 & x \\ 0 & 0 & 1 \end{bmatrix} \begin{bmatrix} c_1 \\ c_2 \\ c_3 \end{bmatrix} \, .$$

$\square$

Remark 1.5. Suppose that $Z' = JZ$ where J is a matrix in JCF but one consisting of several blocks, not just one block. We can see that this systems decomposes into several systems, one corresponding to each block, and that these systems are uncoupled, so we may solve them each separately, using Theorem 1.4, and then simply assemble these individual solutions together to obtain a solution of the general system.

We now illustrate this (confining our illustrations to the case that A is not diagonalizable, as we have already illustrated the diagonalizable case).

Example 1.6.   Consider the system

$$Y' = AY \quad \text{where} \quad A = \begin{bmatrix} 0 & 1 \\ -4 & 4 \end{bmatrix}.$$

We saw in Example 2.12 in Chapter 1 that $A = PJP^{-1}$ with

$$P = \begin{bmatrix} -2 & 1 \\ -4 & 0 \end{bmatrix} \quad \text{and} \quad J = \begin{bmatrix} 2 & 1 \\ 0 & 2 \end{bmatrix}.$$

Then $Z' = JZ$ has solution

$$Z = e^{2x} \begin{bmatrix} 1 & x \\ 0 & 1 \end{bmatrix} \begin{bmatrix} c_1 \\ c_2 \end{bmatrix} = \begin{bmatrix} e^{2x} & xe^{2x} \\ 0 & e^{2x} \end{bmatrix} \begin{bmatrix} c_1 \\ c_2 \end{bmatrix} = M_Z C = \begin{bmatrix} c_1 e^{2x} + c_2 x e^{2x} \\ c_2 e^{2x} \end{bmatrix}$$

and so $Y = PZ = PM_Z C$, i.e.,

$$Y = \begin{bmatrix} -2 & 1 \\ -4 & 0 \end{bmatrix} e^{2x} \begin{bmatrix} 1 & x \\ 0 & 1 \end{bmatrix} \begin{bmatrix} c_1 \\ c_2 \end{bmatrix}$$

$$= \begin{bmatrix} -2 & 1 \\ -4 & 0 \end{bmatrix} \begin{bmatrix} e^{2x} & xe^{2x} \\ 0 & e^{2x} \end{bmatrix} \begin{bmatrix} c_1 \\ c_2 \end{bmatrix}$$

$$= \begin{bmatrix} -2e^{2x} & -2xe^{2x} + e^{2x} \\ -4e^{2x} & -4xe^{2x} \end{bmatrix} \begin{bmatrix} c_1 \\ c_2 \end{bmatrix}$$

$$= \begin{bmatrix} (-2c_1 + c_2)e^{2x} - 2c_2 x e^{2x} \\ -4c_1 e^{2x} - 4c_2 x e^{2x} \end{bmatrix}.$$

Example 1.7.   Consider the system

$$Y' = AY \quad \text{where} \quad A = \begin{bmatrix} 2 & 1 & 1 \\ 2 & 1 & -2 \\ -1 & 0 & -2 \end{bmatrix}.$$

We saw in Example 2.25 in Chapter 1 that $A = PJP^{-1}$ with

$$P = \begin{bmatrix} 1 & 0 & -5 \\ -2 & 0 & -6 \\ -1 & 1 & 1 \end{bmatrix} \quad \text{and } J = \begin{bmatrix} -1 & 1 & 0 \\ 0 & -1 & 0 \\ 0 & 0 & 3 \end{bmatrix}.$$

Then $Z' = JZ$ has solution

$$Z = \begin{bmatrix} e^{-x} & xe^{-x} & 0 \\ 0 & e^{-x} & 0 \\ 0 & 0 & e^{3x} \end{bmatrix} \begin{bmatrix} c_1 \\ c_2 \\ c_3 \end{bmatrix} = M_Z C$$

and so $Y = PZ = PM_Z C$, i.e.,

$$Y = \begin{bmatrix} 1 & 0 & -5 \\ -2 & 0 & -6 \\ -1 & 1 & 1 \end{bmatrix} \begin{bmatrix} e^{-x} & xe^{-x} & 0 \\ 0 & e^{-x} & 0 \\ 0 & 0 & e^{3x} \end{bmatrix} \begin{bmatrix} c_1 \\ c_2 \\ c_3 \end{bmatrix}$$

$$= \begin{bmatrix} e^{-x} & xe^{-x} & -5e^{3x} \\ -2e^{-x} & -2xe^{-x} & -6e^{3x} \\ -e^{-x} & -xe^{-x} + e^{-x} & e^{3x} \end{bmatrix} \begin{bmatrix} c_1 \\ c_2 \\ c_3 \end{bmatrix}$$

$$= \begin{bmatrix} c_1 e^{-x} + c_2 xe^{-x} - 5c_3 e^{3x} \\ -2c_1 e^{-x} - 2c_2 xe^{-x} - 6c_3 e^{3x} \\ (-c_1 + c_2)e^{-x} - c_2 xe^{-x} + c_3 e^{3x} \end{bmatrix}.$$

Example 1.8.   Consider the system

$$Y' = AY \quad \text{where} \quad A = \begin{bmatrix} 2 & 1 & 1 \\ -2 & -1 & -2 \\ 1 & 1 & 2 \end{bmatrix}.$$

We saw in Example 2.26 in Chapter 1 that $A = PJP^{-1}$ with

$$P = \begin{bmatrix} 1 & 1 & 1 \\ -2 & 0 & 0 \\ 1 & 0 & 1 \end{bmatrix} \quad \text{and} \quad J = \begin{bmatrix} 1 & 1 & 0 \\ 0 & 1 & 0 \\ 0 & 0 & 1 \end{bmatrix}.$$

Then $Z' = JZ$ has solution

$$Z = \begin{bmatrix} e^x & xe^x & 0 \\ 0 & e^x & 0 \\ 0 & 0 & e^x \end{bmatrix} \begin{bmatrix} c_1 \\ c_2 \\ c_3 \end{bmatrix} = M_Z C$$

and so $Y = PZ = PM_Z C$, i.e.,

$$Y = \begin{bmatrix} 1 & 1 & 1 \\ -2 & 0 & 0 \\ 1 & 0 & 1 \end{bmatrix} \begin{bmatrix} e^x & xe^x & 0 \\ 0 & e^x & 0 \\ 0 & 0 & e^x \end{bmatrix} \begin{bmatrix} c_1 \\ c_2 \\ c_3 \end{bmatrix}$$

$$= \begin{bmatrix} e^x & xe^x + e^x & e^x \\ -2e^x & -2xe^x & 0 \\ e^x & xe^x & e^x \end{bmatrix} \begin{bmatrix} c_1 \\ c_2 \\ c_3 \end{bmatrix}$$

$$= \begin{bmatrix} (c_1 + c_2 + c_3)e^x + c_2 xe^x \\ -2c_1 e^x & - 2c_2 xe^x \\ (c_1 + c_3)e^x & + c_2 xe^x \end{bmatrix}.$$

Example 1.9.   Consider the system

$$Y' = AY \quad \text{where} \quad A = \begin{bmatrix} 5 & 0 & 1 \\ 1 & 1 & 0 \\ -7 & 1 & 0 \end{bmatrix}.$$

We saw in Example 2.27 in Chapter 1 that $A = PJP^{-1}$ with

$$P = \begin{bmatrix} 2 & 3 & 1 \\ 2 & 1 & 0 \\ -6 & -7 & 0 \end{bmatrix} \quad \text{and} \quad J = \begin{bmatrix} 2 & 1 & 0 \\ 0 & 2 & 1 \\ 0 & 0 & 2 \end{bmatrix}.$$

Then $Z' = JZ$ has solution

$$Z = \begin{bmatrix} e^{2x} & xe^{2x} & (x^2/2)e^{2x} \\ 0 & e^{2x} & xe^{2x} \\ 0 & 0 & e^{2x} \end{bmatrix} \begin{bmatrix} c_1 \\ c_2 \\ c_3 \end{bmatrix} = M_Z C$$

and so $Y = PZ = PM_Z C$, i.e.,

$$Y = \begin{bmatrix} 2 & 3 & 1 \\ 2 & 1 & 0 \\ -6 & -7 & 0 \end{bmatrix} \begin{bmatrix} e^{2x} & xe^{2x} & (x^2/2)e^{2x} \\ 0 & e^{2x} & xe^{2x} \\ 0 & 0 & e^{2x} \end{bmatrix} \begin{bmatrix} c_1 \\ c_2 \\ c_3 \end{bmatrix}$$

$$= \begin{bmatrix} 2e^{2x} & 2xe^{2x} + 3e^{2x} & x^2 e^{2x} + 3xe^{2x} + e^{2x} \\ 2e^{2x} & 2xe^{2x} + e^{2x} & x^2 e^{2x} + xe^{2x} \\ -6e^{2x} & -6xe^{2x} - 7e^{2x} & -3x^2 e^{2x} - 7xe^{2x} \end{bmatrix} \begin{bmatrix} c_1 \\ c_2 \\ c_3 \end{bmatrix}$$

$$= \begin{bmatrix} (2c_1 + 3c_2 + c_3)e^{2x} + (2c_2 + 3c_3)xe^{2x} + c_3 x^2 e^{2x} \\ (2c_1 + c_2)e^{2x} + (2c_2 + c_3)xe^{2x} + c_3 x^2 e^{2x} \\ (-6c_1 - 7c_2)e^{2x} + (-6c_2 - 7c_3)xe^{2x} - 3c_3 x^2 e^{2x} \end{bmatrix}.$$

We conclude this section by showing how to solve initial value problems. This is just one more step, given what we have already done.

Example 1.10.   Consider the initial value problem

$$Y' = AY \quad \text{where} \quad A = \begin{bmatrix} 0 & 1 \\ -4 & 4 \end{bmatrix}, \quad \text{and} \quad Y(0) = \begin{bmatrix} 3 \\ -8 \end{bmatrix}.$$

In Example 1.6, we saw that this system has the general solution

$$Y = \begin{bmatrix} (-2c_1 + c_2)e^{2x} - 2c_2xe^{2x} \\ -4c_1e^{2x} \quad\quad - 4c_2xe^{2x} \end{bmatrix}.$$

Applying the initial condition (i.e., substituting $x = 0$ in this matrix), gives

$$\begin{bmatrix} 3 \\ -8 \end{bmatrix} = Y(0) = \begin{bmatrix} -2c_1 + c_2 \\ -4c_1 \end{bmatrix}$$

with solution

$$\begin{bmatrix} c_1 \\ c_2 \end{bmatrix} = \begin{bmatrix} 2 \\ 7 \end{bmatrix}.$$

Substituting these values in the above matrix gives

$$Y = \begin{bmatrix} 3e^{2x} - 14xe^{2x} \\ -8e^{2x} - 28te^{2x} \end{bmatrix}.$$

Example 1.11.   Consider the initial value problem

$$Y' = AY \quad \text{where} \quad A = \begin{bmatrix} 2 & 1 & 1 \\ 2 & 1 & -2 \\ -1 & 0 & 2 \end{bmatrix}, \quad \text{and} \quad Y(0) = \begin{bmatrix} 8 \\ 32 \\ 5 \end{bmatrix}.$$

In Example 1.8, we saw that this system has the general solution

$$Y = \begin{bmatrix} c_1e^{-x} \quad\quad + c_2xe^{-x} \quad - 5c_3xe^{3x} \\ -2c_1e^{-x} \quad\quad - 2c_2xe^{-x} - 6c_3e^{3x} \\ (-c_1 + c_2)e^{-x} - c_2xe^{-x} + c_3e^{3x} \end{bmatrix}.$$

Applying the initial condition (i.e., substituting $x = 0$ in this matrix) gives

$$\begin{bmatrix} 8 \\ 32 \\ 5 \end{bmatrix} = Y(0) = \begin{bmatrix} c_1 \quad\quad - 5c_3 \\ -2c_1 \quad\quad - 6c_3 \\ -c_1 + c_2 + c_3 \end{bmatrix}$$

with solution

$$\begin{bmatrix} c_1 \\ c_2 \\ c_3 \end{bmatrix} = \begin{bmatrix} -7 \\ 1 \\ -3 \end{bmatrix} .$$

Substituting these values in the above matrix gives

$$Y = \begin{bmatrix} -7e^{-x} + xe^{-x} + 15e^{3x} \\ 14e^{-x} - 2xe^{-x} + 18e^{3x} \\ 8e^{-x} - xe^{-x} - 3e^{3x} \end{bmatrix} .$$

Remark 1.12.   There is a variant on our method of solving systems or initial value problems.

We have written our solution of $Z' = JZ$ as $Z = M_Z C$. Let us be more explicit here and write this solution as

$$Z(x) = M_Z(x)C .$$

This notation reminds us that $Z(x)$ is a vector of functions, $M_Z(x)$ is a matrix of functions, and C is a vector of constants. The key observation is that $M_Z(0) = I$, the identity matrix. Thus, if we wish to solve the initial value problem

$$Z' = JZ, \quad Z(0) = Z_0 ,$$

we find that, in general,

$$Z(x) = M_Z(x)C$$

and, in particular,

$$Z_0 = Z(0) = M_Z(0)C = IC = C ,$$

so the solution to this initial value problem is

$$Z(x) = M_Z(x)Z_0 .$$

Now suppose we wish to solve the system $Y' = AY$. Then, if $A = PJP^{-1}$, we have seen that this system has solution $Y = PZ = PM_ZC$. Let us manipulate this a bit:

$$Y = PM_ZC = PM_ZIC = PM_Z(P^{-1}P)C$$
$$= (PM_ZP^{-1})(PC) .$$

Now let us set $M_Y = PM_ZP^{-1}$, and also let us set $\Gamma = PC$. Note that M_Y is still a matrix of functions, and that Γ is still a vector of arbitrary constants (since P is an invertible constant matrix and C is a vector of arbitrary constants). Thus, with this notation, we see that

$$Y' = AY \qquad has\ solution \qquad Y = M_Y\Gamma .$$

Now suppose we wish to solve the initial value problem

$$Y' = AY, \quad Y(0) = Y_0.$$

Rewriting the above solution of $Y' = AY$ to explicitly include the independent variable, we see that we have

$$Y(x) = M_Y(x)\Gamma$$

and, in particular,

$$Y_0 = Y(0) = M_Y(0)\Gamma = PM_Z(0)P^{-1}\Gamma = PIP^{-1}\Gamma = \Gamma,$$

so we see that

$$Y' = AY, \quad Y(0) = Y_0 \quad \textit{has solution} \quad Y(x) = M_Y(x)Y_0.$$

This variant method has pros and cons. It is actually less effective than our original method for solving a single initial value problem (as it requires us to compute P^{-1} and do some extra matrix multiplication), but it has the advantage of expressing the solution directly in terms of the initial conditions. This makes it more effective if the same system $Y' = AY$ is to be solved for a variety of initial conditions. Also, as we see from Remark 1.14 below, it is of considerable theoretical importance.

Let us now apply this variant method.

Example 1.13.   Consider the initial value problem

$$Y' = AY \quad \text{where} \quad A = \begin{bmatrix} 0 & 1 \\ -4 & 4 \end{bmatrix}, \quad \text{and} \quad Y(0) = \begin{bmatrix} a_1 \\ a_2 \end{bmatrix}.$$

As we have seen in Example 1.6, $A = PJP^{-1}$ with $P = \begin{bmatrix} -2 & 1 \\ -4 & 0 \end{bmatrix}$ and $J = \begin{bmatrix} 2 & 1 \\ 0 & 2 \end{bmatrix}$. Then $M_Z(x) = \begin{bmatrix} e^{2x} & xe^{2x} \\ 0 & e^{2x} \end{bmatrix}$ and

$$\begin{aligned} M_Y(x) = PM_Z(x)P^{-1} &= \begin{bmatrix} -2 & 1 \\ -4 & 0 \end{bmatrix} \begin{bmatrix} e^{2x} & xe^{2x} \\ 0 & e^{2x} \end{bmatrix} \begin{bmatrix} -2 & 1 \\ -4 & 0 \end{bmatrix}^{-1} \\ &= \begin{bmatrix} e^{2x} - 2xe^{2x} & xe^{2x} \\ -4xe^{2x} & e^{2x} + 2xe^{2x} \end{bmatrix} \end{aligned}$$

so

$$\begin{aligned} Y(x) = M_Y(x) \begin{bmatrix} a_1 \\ a_2 \end{bmatrix} &= \begin{bmatrix} e^{2x} - 2xe^{2x} & xe^{2x} \\ -4xe^{2x} & e^{2x} + 2xe^{2x} \end{bmatrix} \begin{bmatrix} a_1 \\ a_2 \end{bmatrix} \\ &= \begin{bmatrix} a_1 e^{2x} + (-2a_1 + a_2)xe^{2x} \\ a_2 e^{2x} + (-4a_1 + 2a_2)xe^{2x} \end{bmatrix}. \end{aligned}$$

In particular, if $Y(0) = \begin{bmatrix} 3 \\ -8 \end{bmatrix}$, then $Y(x) = \begin{bmatrix} 3e^{2x} - 14xe^{2x} \\ -8e^{2x} - 28xe^{2x} \end{bmatrix}$, recovering the result of Example 1.10. But also, if $Y(0) = \begin{bmatrix} 2 \\ 5 \end{bmatrix}$, then $Y(x) = \begin{bmatrix} 2e^{2x} + xe^{2x} \\ 5e^{2x} + 2te^{2x} \end{bmatrix}$, and if $Y(0) = \begin{bmatrix} -4 \\ 15 \end{bmatrix}$, then $Y(x) = \begin{bmatrix} -4e^{2x} + 23xe^{2x} \\ 15e^{2x} + 46xe^{2x} \end{bmatrix}$, etc.

Remark 1.14. In Section 2.4 we will define the matrix exponential, and, with this definition, $M_Z(x) = e^{Jx}$ and $M_Y(x) = PM_Z(x)P^{-1} = e^{Ax}$.

EXERCISES FOR SECTION 2.1

For each exercise, see the corresponding exercise in Chapter 1. In each exercise:

(a) Solve the system $Y' = AY$.

(b) Solve the initial value problem $Y' = AY, Y(0) = Y_0$.

1. $A = \begin{bmatrix} 75 & 56 \\ -90 & -67 \end{bmatrix}$ and $Y_0 = \begin{bmatrix} 1 \\ -1 \end{bmatrix}$.

2. $A = \begin{bmatrix} -50 & 99 \\ -20 & 39 \end{bmatrix}$ and $Y_0 = \begin{bmatrix} 7 \\ 3 \end{bmatrix}$.

3. $A = \begin{bmatrix} -18 & 9 \\ -49 & 24 \end{bmatrix}$ and $Y_0 = \begin{bmatrix} 41 \\ 98 \end{bmatrix}$.

4. $A = \begin{bmatrix} 1 & 1 \\ -16 & 9 \end{bmatrix}$ and $Y_0 = \begin{bmatrix} 7 \\ 16 \end{bmatrix}$.

5. $A = \begin{bmatrix} 2 & 1 \\ -25 & 12 \end{bmatrix}$ and $Y_0 = \begin{bmatrix} -10 \\ -75 \end{bmatrix}$.

6. $A = \begin{bmatrix} -15 & 9 \\ -25 & 15 \end{bmatrix}$ and $Y_0 = \begin{bmatrix} 50 \\ 100 \end{bmatrix}$.

7. $A = \begin{bmatrix} 1 & 0 & 0 \\ 1 & 2 & -3 \\ 1 & -1 & 0 \end{bmatrix}$ and $Y_0 = \begin{bmatrix} 6 \\ -10 \\ 10 \end{bmatrix}$.

8. $A = \begin{bmatrix} 3 & 0 & 2 \\ 1 & 3 & 1 \\ 0 & 1 & 1 \end{bmatrix}$ and $Y_0 = \begin{bmatrix} 0 \\ 3 \\ 3 \end{bmatrix}$.

9. $A = \begin{bmatrix} 5 & 8 & 16 \\ 4 & 1 & 8 \\ -4 & -4 & -11 \end{bmatrix}$ and $Y_0 = \begin{bmatrix} 0 \\ 2 \\ -1 \end{bmatrix}$.

10. $A = \begin{bmatrix} 4 & 2 & 3 \\ -1 & 1 & -3 \\ 2 & 4 & 9 \end{bmatrix}$ and $Y_0 = \begin{bmatrix} 3 \\ 2 \\ 1 \end{bmatrix}$.

11. $A = \begin{bmatrix} 5 & 2 & 1 \\ -1 & 2 & -1 \\ -1 & -2 & 3 \end{bmatrix}$ and $Y_0 = \begin{bmatrix} -3 \\ 2 \\ 9 \end{bmatrix}$.

12. $A = \begin{bmatrix} 8 & -3 & -3 \\ 4 & 0 & -2 \\ -2 & 1 & 3 \end{bmatrix}$ and $Y_0 = \begin{bmatrix} 5 \\ 8 \\ 7 \end{bmatrix}$.

13. $A = \begin{bmatrix} -3 & 1 & -1 \\ -7 & 5 & -1 \\ -6 & 6 & -2 \end{bmatrix}$ and $Y_0 = \begin{bmatrix} -1 \\ 3 \\ 6 \end{bmatrix}$.

14. $A = \begin{bmatrix} 3 & 0 & 0 \\ 9 & -5 & -18 \\ -4 & 4 & 12 \end{bmatrix}$ and $Y_0 = \begin{bmatrix} 2 \\ -1 \\ 1 \end{bmatrix}$.

15. $A = \begin{bmatrix} -6 & 9 & 0 \\ -6 & 6 & -2 \\ 9 & -9 & 3 \end{bmatrix}$ and $Y_0 = \begin{bmatrix} 1 \\ 3 \\ -6 \end{bmatrix}$.

16. $A = \begin{bmatrix} -18 & 42 & 168 \\ 1 & -7 & -40 \\ -2 & 6 & 27 \end{bmatrix}$ and $Y_0 = \begin{bmatrix} 7 \\ -2 \\ 1 \end{bmatrix}$.

17. $A = \begin{bmatrix} -1 & 1 & -1 \\ -10 & 6 & -5 \\ -6 & 3 & 2 \end{bmatrix}$ and $Y_0 = \begin{bmatrix} 3 \\ 10 \\ 18 \end{bmatrix}$.

18. $A = \begin{bmatrix} 0 & -4 & 1 \\ 2 & -6 & 1 \\ 4 & -8 & 0 \end{bmatrix}$ and $Y_0 = \begin{bmatrix} 2 \\ 5 \\ 8 \end{bmatrix}$.

19. $A = \begin{bmatrix} -4 & 1 & 2 \\ -5 & 1 & 3 \\ -7 & 2 & 3 \end{bmatrix}$ and $Y_0 = \begin{bmatrix} 6 \\ 11 \\ 9 \end{bmatrix}$.

20. $A = \begin{bmatrix} -4 & -2 & 5 \\ -1 & -1 & 1 \\ -2 & -1 & 2 \end{bmatrix}$ and $Y_0 = \begin{bmatrix} 9 \\ 5 \\ 8 \end{bmatrix}$.

2.2 HOMOGENEOUS SYSTEMS WITH CONSTANT COEFFICIENTS: COMPLEX ROOTS

In this section, we show how to solve a homogeneous system $Y' = AY$ where the characteristic polynomial of A has complex roots. In principle, this is the same as the situation where the characteristic polynomial of A has real roots, which we dealt with in Section 2.1, but in practice, there is an extra step in the solution.

We will begin by doing an example, which will show us where the difficulty lies, and then we will overcome that difficulty. But first, we need some background.

Definition 2.1.   For a complex number z, the exponential e^z is defined by

$$e^z = 1 + z + z^2/2! + z^3/3! + \dots .$$

The complex exponential has the following properties.

Theorem 2.2.   *(1) (Euler) For any θ,*

$$e^{i\theta} = \cos(\theta) + i\sin(\theta) .$$

(2) For any a,

$$\frac{d}{dz}(e^{az}) = ae^{az} .$$

(3) For any z_1 and z_2,

$$e^{z_1+z_2} = e^{z_1}e^{z_2} .$$

(4) If $z = s + it$, then

$$e^z = e^s(\cos(t) + i\sin(t)) .$$

(5) For any z,

$$e^{\bar{z}} = \overline{e^z} .$$

Proof. For the proof, see Theorem 2.2 in Appendix A.   □

The following lemma will save us some computations.

Lemma 2.3.   *Let A be a matrix with real entries, and let v be an eigenvector of A with associated eigenvalue λ. Then $\bar{v}$ is an eigenvector of A with associated eigenvalue $\bar{\lambda}$.*

Proof. We have that $Av = \lambda v$, by hypothesis. Let us take the complex conjugate of each side of this equation. Then

$$\overline{Av} = \overline{\lambda v},$$
$$\overline{A}\,\overline{v} = \overline{\lambda}\,\overline{v},$$
$$A\bar{v} = \bar{\lambda}\bar{v} \text{ (as } \overline{A} = A \text{ since all the entries of } A \text{ are real) },$$

as claimed.   □

Now for our example.

Example 2.4.   Consider the system

$$Y' = AY \quad \text{where} \quad A = \begin{bmatrix} 2 & -17 \\ 1 & 4 \end{bmatrix}.$$

A has characteristic polynomial $\lambda^2 - 6\lambda + 25$ with roots $\lambda_1 = 3 + 4i$ and $\lambda_2 = \overline{\lambda_1} = 3 - 4i$, each of multiplicity 1. Thus, λ_1 and λ_2 are the eigenvalues of A, and we compute that the eigenspace $E_{3+4i} = \text{Ker}(A - (3+4i)I)$ has basis $\left\{ v_1 = \begin{bmatrix} -1 + 4i \\ 1 \end{bmatrix} \right\}$, and hence, by Lemma 2.3, that the eigenspace $E_{3-4i} = \text{Ker}(A - (3-4i)I)$ has basis $\left\{ v_2 = \overline{v_1} = \begin{bmatrix} -1 - 4i \\ 1 \end{bmatrix} \right\}$. Hence, just as before,

$$A = PJP^{-1} \text{ with } P = \begin{bmatrix} -1 + 4i & -1 - 4i \\ 1 & 1 \end{bmatrix} \text{ and } J = \begin{bmatrix} 3 + 4i & 0 \\ 0 & 3 - 4i \end{bmatrix}.$$

We continue as before, but now we use F to denote a vector of arbitrary constants. (This is just for neatness. Our constants will change, as you will see, and we will use the vector C to denote our final constants, as usual.) Then $Z' = JZ$ has solution

$$Z = \begin{bmatrix} e^{(3+4i)x} & 0 \\ 0 & e^{(3-4i)x} \end{bmatrix} \begin{bmatrix} f_1 \\ f_2 \end{bmatrix} = M_Z F = \begin{bmatrix} f_1 e^{(3+4i)x} \\ f_2 e^{(3-4i)x} \end{bmatrix}$$

and so $Y = PZ = PM_Z F$, i.e.,

$$Y = \begin{bmatrix} -1 + 4i & -1 - 4i \\ 1 & 1 \end{bmatrix} \begin{bmatrix} e^{(3+4i)x} & 0 \\ 0 & e^{(3-4i)x} \end{bmatrix} \begin{bmatrix} f_1 \\ f_2 \end{bmatrix}$$

$$= f_1 e^{(3+4i)x} \begin{bmatrix} -1 + 4i \\ 1 \end{bmatrix} + f_2 e^{(3-4i)x} \begin{bmatrix} -1 - 4i \\ 1 \end{bmatrix}.$$

Now we want our differential equation to have real solutions, and in order for this to be the case, it turns out that we must have $f_2 = \overline{f_1}$. Thus, we may write our solution as

$$Y = f_1 e^{(3+4i)x} \begin{bmatrix} -1 + 4i \\ 1 \end{bmatrix} + \overline{f_1} e^{(3-4i)x} \begin{bmatrix} -1 - 4i \\ 1 \end{bmatrix}$$

$$= f_1 e^{(3+4i)x} \begin{bmatrix} -1 + 4i \\ 1 \end{bmatrix} + \overline{f_1 e^{(3+4i)x} \begin{bmatrix} -1 + 4i \\ 1 \end{bmatrix}},$$

where f_1 is an arbitrary complex constant.

This solution is correct but unacceptable. We want to solve the system $Y' = AY$, where A has real coefficients, and we have a solution which is indeed a real vector, but this vector is expressed in terms of complex numbers and functions. We need to obtain a solution that is expressed totally in terms of real numbers and functions. In order to do this, we need an extra step.

In order not to interrupt the flow of exposition, we simply state here what we need to do, and we justify this after the conclusion of the example.

We therefore do the following: We simply replace the matrix PM_Z by the matrix whose first column is the real part $\text{Re}(e^{\lambda_1 x} v_1) = \text{Re}\left(e^{(3+4i)x} \begin{bmatrix} -1 + 4i \\ 1 \end{bmatrix} \right)$, and whose second column is the imaginary part $\text{Im}(e^{\lambda_1 x} v_1) = \text{Im}\left(e^{(3+4i)x} \begin{bmatrix} -1 + 4i \\ 1 \end{bmatrix} \right)$, and the vector F by the vector C of arbitrary real constants. We compute

$$
e^{(3+4i)x} \begin{bmatrix} -1 + 4i \\ 1 \end{bmatrix} = e^{3x} (\cos(4x) + i \sin(4x)) \begin{bmatrix} -1 + 4i \\ 1 \end{bmatrix}
$$
$$
= e^{3x} \begin{bmatrix} -\cos(4x) - 4\sin(4x) \\ \cos(4x) \end{bmatrix} + i e^{3x} \begin{bmatrix} 4\cos(4x) - \sin(4x) \\ \sin(4x) \end{bmatrix}
$$

and so we obtain

$$
Y = \begin{bmatrix} e^{3x}(-\cos(4x) - 4\sin(4x)) & e^{3x}(4\cos(4x) - \sin(4x)) \\ e^{3x}\cos(4x) & e^{3x}\sin(4x) \end{bmatrix} \begin{bmatrix} c_1 \\ c_2 \end{bmatrix}
$$
$$
= \begin{bmatrix} (-c_1 + 4c_2)e^{3x}\cos(4x) + (-4c_1 - c_2)e^{3x}\sin(4x) \\ c_1 e^{3x}\cos(4x) + c_2 e^{3x}\sin(4x) \end{bmatrix}.
$$

Now we justify the step we have done.

Lemma 2.5. *Consider the system $Y' = AY$, where A is a matrix with real entries. Let this system have general solution of the form*

$$
Y = PM_Z F = \begin{bmatrix} v_1 \mid \overline{v_1} \end{bmatrix} \begin{bmatrix} e^{\lambda_1 x} & 0 \\ 0 & e^{\overline{\lambda_1} x} \end{bmatrix} \begin{bmatrix} f_1 \\ \overline{f_1} \end{bmatrix} = \begin{bmatrix} e^{\lambda_1 x} v_1 \mid \overline{e^{\lambda_1 x} v_1} \end{bmatrix} \begin{bmatrix} f_1 \\ \overline{f_1} \end{bmatrix},
$$

where f_1 is an arbitrary complex constant. Then this system also has general solution of the form

$$
Y = \begin{bmatrix} \text{Re}(e^{\lambda_1 x} v_1) \mid \text{Im}(e^{\lambda_1 x} v_1) \end{bmatrix} \begin{bmatrix} c_1 \\ c_2 \end{bmatrix},
$$

where c_1 and c_2 are arbitrary real constants.

Proof. First note that for any complex number $z = x + iy$, $x = \text{Re}(z) = \frac{1}{2}(z + \overline{z})$ and $y = \text{Im}(z) = \frac{1}{2i}(z - \overline{z})$, and similarly, for any complex vector.

Now $Y' = AY$ has general solution $Y = PM_Z F = PM_Z(RR^{-1})F = (PM_Z R)(R^{-1}F)$ for any invertible matrix R. We now (cleverly) choose

$$R = \begin{bmatrix} 1/2 & 1/(2i) \\ 1/2 & -1/(2i) \end{bmatrix} .$$

With this choice of R,

$$PM_Z R = \left[\operatorname{Re}(e^{\lambda_1 x}v_1) \,\middle|\, \operatorname{Im}(e^{\lambda_1 x}v_1) \right] .$$

Then

$$R^{-1} = \begin{bmatrix} 1 & 1 \\ i & -i \end{bmatrix} .$$

Since f_1 is an arbitrary complex constant, we may (cleverly) choose to write it as $f_1 = \frac{1}{2}(c_1 + ic_2)$ for arbitrary real constants c_1 and c_2, and with this choice

$$R^{-1}F = \begin{bmatrix} c_1 \\ c_2 \end{bmatrix} ,$$

yielding a general solution as claimed.   □

We now solve $Y' = AY$ where A is a real 3-by-3 matrix with a pair of complex eigenvalues and a third, real eigenvalue. As you will see, we use the idea of Lemma 2.5 to simply replace the "relevant" columns of PM_Z in order to obtain our final solution.

Example 2.6.   Consider the system

$$Y' = AY \quad \text{where} \quad A = \begin{bmatrix} 15 & -16 & 8 \\ 10 & -10 & 5 \\ 0 & 1 & 2 \end{bmatrix} .$$

A has characteristic polynomial $(\lambda^2 - 2\lambda + 5)(\lambda - 5)$ with roots $\lambda_1 = 1 + 2i$, $\lambda_2 = \overline{\lambda_1} = 1 - 2i$, and $\lambda_3 = 5$, each of multiplicity 1. Thus, λ_1, λ_2, and λ_3 are the eigenvalues of A, and we compute that the eigenspace $E_{1+2i} = \operatorname{Ker}(A - (1 + 2i)I)$ has basis $\left\{ v_1 = \begin{bmatrix} -2 + 2i \\ -1 + 2i \\ 1 \end{bmatrix} \right\}$, and hence, by

Lemma 2.3, that the eigenspace $E_{1-2i} = \operatorname{Ker}(A - (1 - 2i)I)$ has basis $\left\{ v_2 = \overline{v_1} = \begin{bmatrix} -2 - 2i \\ -1 - 2i \\ 1 \end{bmatrix} \right\}$.

We further compute that the eigenspace $E_5 = \operatorname{Ker}(A - 5I)$ has basis $\left\{ v_3 = \begin{bmatrix} 4 \\ 3 \\ 1 \end{bmatrix} \right\}$. Hence, just as before,

$$A = PJP^{-1} \text{ with } P = \begin{bmatrix} -2+2i & -2-2i & 4 \\ -1+2i & -1-2i & 3 \\ 1 & 1 & 1 \end{bmatrix} \text{ and } J = \begin{bmatrix} 1+2i & 0 & 0 \\ 0 & 1-2i & 0 \\ 0 & 0 & 5 \end{bmatrix}.$$

Then $Z' = JZ$ has solution

$$Z = \begin{bmatrix} e^{(1+2i)x} & 0 & 0 \\ 0 & e^{(1-2i)x} & 0 \\ 0 & 0 & e^{5x} \end{bmatrix} \begin{bmatrix} f_1 \\ f_1 \\ c_3 \end{bmatrix} = M_Z F = \begin{bmatrix} f_1 e^{(1+2i)x} \\ f_1 e^{(1+2i)x} \\ c_3 e^{5x} \end{bmatrix}$$

and so $Y = PZ = PM_Z F$, i.e.,

$$Y = \begin{bmatrix} -2+2i & -2-2i & 4 \\ -1+2i & -1-2i & 3 \\ 1 & 1 & 1 \end{bmatrix} \begin{bmatrix} e^{(1+2i)x} & 0 & 0 \\ 0 & e^{(1-2i)x} & 0 \\ 0 & 0 & e^{5x} \end{bmatrix} \begin{bmatrix} f_1 \\ f_1 \\ c_3 \end{bmatrix}.$$

Now

$$e^{(1+2i)x} \begin{bmatrix} -2+2i \\ -1+2i \\ 1 \end{bmatrix} = e^x(\cos(2x) + i\sin(2x)) \begin{bmatrix} -2+2i \\ -1+2i \\ 1 \end{bmatrix}$$

$$= \begin{bmatrix} e^x(-2\cos(2x) - 2\sin(2x)) \\ e^x(-\cos(2x) - 2\sin(2x)) \\ e^x\cos(2x) \end{bmatrix} + i \begin{bmatrix} e^x(2\cos(2x) - 2\sin(2x)) \\ e^x(2\cos(2x) - \sin(2x)) \\ e^x\sin(2x) \end{bmatrix}$$

and of course

$$e^{5x} \begin{bmatrix} 4 \\ 3 \\ 1 \end{bmatrix} = \begin{bmatrix} 4e^{5x} \\ 3e^{5x} \\ e^{5x} \end{bmatrix},$$

so, replacing the relevant columns of PM_Z, we find

$$Y = \begin{bmatrix} e^x(-2\cos(2x) - 2\sin(2x)) & e^x(2\cos(2x) - 2\sin(2x)) & 4e^{5x} \\ e^x(-\cos(2x) - 2\sin(2x)) & e^x(2\cos(2x) - \sin(2x)) & 3e^{5x} \\ e^x\cos(2x) & e^x\sin(2x) & e^{5x} \end{bmatrix} \begin{bmatrix} c_1 \\ c_2 \\ c_3 \end{bmatrix}$$

$$= \begin{bmatrix} (-2c_1 + 2c_2)e^x\cos(2x) + (-2c_1 - 2c_2)e^x\sin(2x) + 4c_3e^{5x} \\ (-c_1 + 2c_2)e^x\cos(2x) + (-2c_1 - c_2)e^x\sin(2x) + 3c_3e^{5x} \\ c_1e^x\cos(2x) + c_2e^x\sin(2x) + c_3e^{5x} \end{bmatrix}.$$

EXERCISES FOR SECTION 2.2

In Exercises 1–4:

(a) Solve the system $Y' = AY$.

(b) Solve the initial value problem $Y' = AY, Y(0) = Y_0$.

In Exercises 5 and 6, solve the system $Y' = AY$.

1. $A = \begin{bmatrix} 3 & 5 \\ -2 & 5 \end{bmatrix}$, $\quad \det(\lambda I - A) = \lambda^2 - 8\lambda + 25$, $\quad$ and $Y_0 = \begin{bmatrix} 8 \\ 13 \end{bmatrix}$.

2. $A = \begin{bmatrix} 3 & 4 \\ -2 & 7 \end{bmatrix}$, $\quad \det(\lambda I - A) = \lambda^2 - 10\lambda + 29$, $\quad$ and $Y_0 = \begin{bmatrix} 3 \\ 5 \end{bmatrix}$.

3. $A = \begin{bmatrix} 5 & 13 \\ -1 & 9 \end{bmatrix}$, $\quad \det(\lambda I - A) = \lambda^2 - 14\lambda + 58$, $\quad$ and $Y_0 = \begin{bmatrix} 2 \\ 1 \end{bmatrix}$.

4. $A = \begin{bmatrix} 7 & 17 \\ -4 & 11 \end{bmatrix}$, $\quad \det(\lambda I - A) = \lambda^2 - 18\lambda + 145$, $\quad$ and $Y_0 = \begin{bmatrix} 5 \\ 2 \end{bmatrix}$.

5. $A = \begin{bmatrix} 37 & 10 & 20 \\ -59 & -9 & -24 \\ -33 & -12 & -21 \end{bmatrix}$, $\quad \det(\lambda I - A) = (\lambda^2 - 4\lambda + 29)(\lambda - 3)$.

6. $A = \begin{bmatrix} -4 & -42 & 15 \\ 4 & 25 & -10 \\ 6 & 32 & -13 \end{bmatrix}$, $\quad \det(\lambda I - A) = (\lambda^2 - 6\lambda + 13)(\lambda - 2)$.

2.3 INHOMOGENEOUS SYSTEMS WITH CONSTANT COEFFICIENTS

In this section, we show how to solve an inhomogeneous system $Y' = AY + G(x)$ where $G(x)$ is a vector of functions. (We will often abbreviate $G(x)$ by G). We use a method that is a direct generalization of the method we used for solving a homogeneous system in Section 2.1.

Consider the matrix system

$$Y' = AY + G.$$

Step 1. Write $A = PJP^{-1}$ with J in JCF, so the system becomes

$$Y' = (PJP^{-1})Y + G$$
$$Y' = PJ(P^{-1}Y) + G$$
$$P^{-1}Y' = J(P^{-1}Y) + P^{-1}G$$
$$(P^{-1}Y)' = J(P^{-1}Y) + P^{-1}G.$$

(Note that, since P^{-1} is a constant matrix, we have that $(P^{-1}Y)' = P^{-1}Y'$.)

Step 2. Set $Z = P^{-1}Y$ and $H = P^{-1}G$, so this system becomes

$$Z' = JZ + H$$

and solve this system for Z.

Step 3. Since $Z = P^{-1}Y$, we have that

$$Y = PZ$$

is the solution to our original system.

Again, the key to this method is to be able to perform Step 2, and again this is straightforward. Within each Jordan block, we solve from the bottom up. Let us focus our attention on a single k-by-k block. The equation for the last function z_k in that block is an inhomogeneous first-order differential equation involving only z_k, and we go ahead and solve it. The equation for the next to the last function z_{k-1} in that block is an inhomogeneous first-order differential equation involving only z_{k-1} and z_k. We substitute in our solution for z_k to obtain an inhomogeneous first-order differential equation for z_{k-1} involving only z_{k-1}, and we go ahead and solve it, etc.

In principle, this is the method we use. In practice, using this method directly is solving each system "by hand," and instead we choose to "automate" this procedure. This leads us to the following method. In order to develop this method we must begin with some preliminaries.

For a fixed matrix A, we say that the inhomogeneous system $Y' = AY + G(x)$ has *associated homogeneous system* $Y' = AY$. By our previous work, we know how to find the general solution of $Y' = AY$. First we shall see that, in order to find the general solution of $Y' = AY + G(x)$, it suffices to find a single solution of that system.

Lemma 3.1. *Let Y_i be any solution of $Y' = AY + G(x)$. If Y_h is any solution of the associated homogeneous system $Y' = AY$, then $Y_h + Y_i$ is also a solution of $Y' = AY + G(x)$, and every solution of $Y' = AY + G(x)$ is of this form.*

Consequently, the general solution of $Y' = AY + G(x)$ is given by $Y = Y_H + Y_i$, where Y_H denotes the general solution of $Y' = AY$.

Proof. First we check that $Y = Y_h + Y_i$ is a solution of $Y' = AY + G(x)$. We simply compute

$$Y' = (Y_h + Y_i)' = Y_h' + Y_i' = (AY_h) + (AY_i + G)$$
$$= A(Y_h + Y_i) + G = AY + G$$

as claimed.

Now we check that every solution Y of $Y' = AY + G(x)$ is of this form. So let Y be any solution of this inhomogeneous system. We can certainly write $Y = (Y - Y_i) + Y_i = Y_h + Y_i$ where $Y_h = Y - Y_i$. We need to show that Y_h defined in this way is indeed a solution of $Y' = AY$. Again we compute

$$Y_h' = (Y - Y_i)' = Y' - Y_i' = (AY + G) - (AY_i + G)$$
$$= A(Y - Y_i) = AY_h$$

as claimed.   □

(It is common to call Y_i a *particular solution* of the inhomogeneous system.)

Let us now recall our work from Section 2.1, and keep our previous notation. The homogeneous system $Y' = AY$ has general solution $Y_H = PM_ZC$ where C is a vector of arbitrary constants. Let us set $N_Y = N_Y(x) = PM_Z(x)$ for convenience, so $Y_H = N_YC$. Then $Y_H' = (N_YC)' = N_Y'C$, and then, substituting in the equation $Y' = AY$, we obtain the equation $N_Y'C = AN_YC$. Since this equation must hold for any C, we conclude that

$$N_Y' = AN_Y .$$

We use this fact to write down a solution to $Y' = AY + G$. We will verify by direct computation that the function we write down is indeed a solution. This verification is not a difficult one, but nevertheless it is a fair question to ask how we came up with this function. Actually, it can be derived in a very natural way, but the explanation for this involves the matrix exponential and so we defer it until Section 2.4. Nevertheless, once we have this solution (no matter how we came up with it) we are certainly free to use it.

It is convenient to introduce the following nonstandard notation. For a vector $H(x)$, we let $\int_0 H(x)dx$ denote an arbitrary but fixed antiderivative of $H(x)$. In other words, in obtaining $\int_0 H(x)dx$, we simply ignore the constants of integration. This is legitimate for our purposes, as by Lemma 3.1 we only need to find a single solution to an inhomogeneous system, and it doesn't matter which one we find—any one will do. (Otherwise said, we can "absorb" the constants of integration into the general solution of the associated homogeneous system.)

Theorem 3.2.   *The function*

$$Y_i = N_Y \int_0 N_Y^{-1} G \, dx$$

is a solution of the system $Y' = AY + G$.

Proof. We simply compute Y_i'. We have

$$Y_i' = \left(N_Y \int_0 N_Y^{-1} G \, dx \right)'$$

$$= N_Y' \int N_Y^{-1} G \, dx + N_Y \left(\int_0 N_Y^{-1} G \, dx \right)'$$

by the product rule

$$= N_Y' \int_0 N_Y^{-1} G \, dx + N_Y (N_Y^{-1} G)$$

by the definition of the antiderivative

$$= N_Y' \int_0 N_Y^{-1} G \, dx + G$$

$$= (A N_Y) \int_0 N_Y^{-1} G \, dx + G$$

as $N_Y' = A N_Y$

$$= A \left(N_Y \int_0 N_Y^{-1} G \, dx \right) + G$$

$$= A Y_i + G$$

as claimed. □

We now do a variety of examples: a 2-by-2 diagonalizable system, a 2-by-2 nondiagonalizable system, a 3-by-3 diagonalizable system, and a 2-by-2 system in which the characteristic polynomial has complex roots. In all these examples, when it comes to finding N_Y^{-1}, it is convenient to use the fact that $N_Y^{-1} = (P M_Z)^{-1} = M_Z^{-1} P^{-1}$.

Example 3.3.   Consider the system

$$Y' = AY + G \quad \text{where} \quad A = \begin{bmatrix} 5 & -7 \\ 2 & -4 \end{bmatrix} \text{ and } G = \begin{bmatrix} 30e^x \\ 60e^{2x} \end{bmatrix}.$$

We saw in Example 1.2 that

$$P = \begin{bmatrix} 7 & 1 \\ 2 & 1 \end{bmatrix} \quad \text{and} \quad M_Z = \begin{bmatrix} e^{3x} & 0 \\ 0 & e^{-2x} \end{bmatrix},$$

and $N_Y = P M_Z$. Then

$$N_Y^{-1} G = \begin{bmatrix} e^{-3x} & 0 \\ 0 & e^{2x} \end{bmatrix} (1/5) \begin{bmatrix} 1 & -1 \\ -2 & 7 \end{bmatrix} \begin{bmatrix} 30e^x \\ 60e^{2x} \end{bmatrix} = \begin{bmatrix} 6e^{-2x} - 12e^{-x} \\ -12e^{3x} + 84e^{4x} \end{bmatrix}.$$

Then

$$\int_0 N_Y^{-1} G = \begin{bmatrix} -3e^{-2x} + 12e^{2x} \\ -4e^{3x} + 21e^{4x} \end{bmatrix}$$

and

$$Y_i = N_Y \int_0^{} N_Y^{-1} G = \begin{bmatrix} 7 & 1 \\ 2 & 1 \end{bmatrix} \begin{bmatrix} e^{3x} & 0 \\ 0 & e^{-2x} \end{bmatrix} \begin{bmatrix} -3e^{-2x} + 12e^{2x} \\ -4e^{3x} + 21e^{4x} \end{bmatrix}$$
$$= \begin{bmatrix} -25e^x + 105e^{2x} \\ -10e^x + 45e^{2x} \end{bmatrix}.$$

Example 3.4.   Consider the system

$$Y' = AY + G \quad \text{where} \quad A = \begin{bmatrix} 0 & 1 \\ -4 & 4 \end{bmatrix} \text{ and } G = \begin{bmatrix} 60e^{3x} \\ 72e^{5x} \end{bmatrix}.$$

We saw in Example 1.6 that

$$P = \begin{bmatrix} -2 & 1 \\ 4 & 0 \end{bmatrix} \quad \text{and} \quad M_Z = e^{2x} \begin{bmatrix} 1 & x \\ 0 & 1 \end{bmatrix},$$

and $N_Y = P M_Z$. Then

$$N_Y^{-1} G = e^{-2x} \begin{bmatrix} 1 & -x \\ 0 & 1 \end{bmatrix} (1/4) \begin{bmatrix} 0 & -1 \\ 4 & -2 \end{bmatrix} \begin{bmatrix} 60e^{3x} \\ 72e^{5x} \end{bmatrix} = \begin{bmatrix} -18e^{3x} - 60xe^x + 36xe^{3x} \\ 60e^x - 36e^{3x} \end{bmatrix}.$$

Then

$$\int_0^{} N_Y^{-1} G = \begin{bmatrix} 60e^x - 60xe^x - 10e^{3x} + 12xe^{3x} \\ 60e^x - 12e^{3x} \end{bmatrix}$$

and

$$Y_i = N_Y \int_0^{} N_Y^{-1} G = \begin{bmatrix} -2 & 1 \\ -4 & 0 \end{bmatrix} e^{2x} \begin{bmatrix} 1 & x \\ 0 & 1 \end{bmatrix} \begin{bmatrix} 60e^x - 60xe^x - 10e^{3x} + 12xe^{3x} \\ 60e^x - 12e^{3x} \end{bmatrix}$$
$$= \begin{bmatrix} -60e^{3x} + 8e^{5x} \\ -240e^{3x} + 40e^{5x} \end{bmatrix}.$$

Example 3.5.   Consider the system

$$Y' = AY + G \quad \text{where} \quad A = \begin{bmatrix} 2 & -3 & -3 \\ 2 & -2 & -2 \\ -2 & 1 & 1 \end{bmatrix} \text{ and } G = \begin{bmatrix} e^x \\ 12e^{3x} \\ 20e^{4x} \end{bmatrix}.$$

We saw in Example 1.3 that

$$P = \begin{bmatrix} 1 & 0 & -1 \\ 0 & -1 & -1 \\ 1 & 1 & 1 \end{bmatrix} \quad \text{and} \quad M_Z = \begin{bmatrix} e^{-x} & 0 & 0 \\ 0 & 1 & 0 \\ 0 & 0 & e^{2x} \end{bmatrix},$$

and $N_Y = P M_Z$. Then

$$N_Y^{-1}G = \begin{bmatrix} e^x & 0 & 0 \\ 0 & 1 & 0 \\ 0 & 0 & e^{-2x} \end{bmatrix} \begin{bmatrix} 0 & 1 & 1 \\ 1 & -2 & -1 \\ -1 & 1 & 1 \end{bmatrix} \begin{bmatrix} e^x \\ 12e^{3x} \\ 20e^{4x} \end{bmatrix} = \begin{bmatrix} 12e^{4x} + 20e^{5x} \\ e^x - 24e^{3x} - 20e^{4x} \\ -e^{-x} + 12e^x + 20e^{2x} \end{bmatrix}.$$

Then

$$\int_0 N_Y^{-1}G = \begin{bmatrix} 3e^{4x} + 4e^{5x} \\ e^x - 8e^{3x} - 5e^{4x} \\ e^{-x} + 12e^x + 10e^{2x} \end{bmatrix}$$

and

$$Y_i = N_Y \int_0 N_Y^{-1}G = \begin{bmatrix} 1 & 0 & -1 \\ 0 & -1 & -1 \\ 1 & 1 & 1 \end{bmatrix} \begin{bmatrix} e^{-x} & 0 & 0 \\ 0 & 1 & 0 \\ 0 & 0 & e^{2x} \end{bmatrix} \begin{bmatrix} 3e^{4x} + 4e^{5x} \\ e^x - 8e^{3x} - 5e^{4x} \\ e^{-x} + 12e^x + 10e^{2x} \end{bmatrix}$$

$$= \begin{bmatrix} -e^x - 9e^{3x} - 6e^{4x} \\ -2e^x - 4e^{3x} - 5e^{4x} \\ 2e^x + 7e^{3x} + 9e^{4x} \end{bmatrix}.$$

Example 3.6. Consider the system

$$Y' = AY + G \quad \text{where} \quad A = \begin{bmatrix} 2 & -17 \\ 1 & 4 \end{bmatrix} \text{ and } G = \begin{bmatrix} 200 \\ 160e^x \end{bmatrix}.$$

We saw in Example 2.4 that

$$P = \begin{bmatrix} -1 + 4i & -1 - 4i \\ 1 & 1 \end{bmatrix} \quad \text{and} \quad M_Z = \begin{bmatrix} e^{(3+4i)x} & 0 \\ 0 & e^{(3-4i)x} \end{bmatrix},$$

and $N_Y = P M_Z$. Then

$$N_Y^{-1}G = \begin{bmatrix} e^{-(3+4i)x} & 0 \\ 0 & e^{-(3-4i)x} \end{bmatrix} (1/(8i)) \begin{bmatrix} 1 & 1 + 4i \\ 1 & 1 - 4i \end{bmatrix} \begin{bmatrix} 200 \\ 160e^x \end{bmatrix}$$

$$= \begin{bmatrix} -25e^{(-3-4i)x} + 20(4 - i)e^{(-2-4i)x} \\ 25e^{(-3+4i)x} + 20(4 + i)e^{(-2+4i)x} \end{bmatrix}.$$

Then

$$\int_0^{} N_Y^{-1}G = \begin{bmatrix} (4+3i)e^{(-3-4i)x} + (-4+18i)e^{(-2-4i)x} \\ (4-3i)e^{(-3+4i)x} + (-4-18i)e^{(-2+4i)x} \end{bmatrix}$$

and

$$Y_i = N_Y \int_0^{} N_Y^{-1}G$$

$$= \begin{bmatrix} -1+4i & -1-4i \\ 1 & 1 \end{bmatrix} \begin{bmatrix} e^{(3+4i)x} & 0 \\ 0 & e^{(3-4i)x} \end{bmatrix} \begin{bmatrix} (4+3i)e^{(-3-4i)x} + (-4+18i)e^{(-2-4i)x} \\ (4-3i)e^{(-3+4i)x} + (-4-18i)e^{(-2+4i)x} \end{bmatrix}$$

$$= \begin{bmatrix} -1+4i & -1-4i \\ 1 & 1 \end{bmatrix} \begin{bmatrix} (4+3i) + (-4+18i)e^x \\ (4-3i) + (-4-18i)e^x \end{bmatrix}$$

$$= \begin{bmatrix} -32 - 136e^x \\ 8 - 8e^x \end{bmatrix}.$$

(Note that in this last example we could do arithmetic with complex numbers directly, i.e., without having to convert complex exponentials into real terms.)

Once we have done this work, it is straightforward to solve initial value problems. We do a single example that illustrates this.

Example 3.7. Consider the initial value problem

$$Y' = AY + G, \quad Y(0) = \begin{bmatrix} 7 \\ 17 \end{bmatrix}, \quad \text{where} \quad A = \begin{bmatrix} 5 & -7 \\ 2 & -4 \end{bmatrix} \text{ and } G = \begin{bmatrix} 30e^x \\ 60e^{2x} \end{bmatrix}.$$

We saw in Example 1.2 that the associated homogenous system has general solution

$$Y_H = \begin{bmatrix} 7c_1 e^{3x} + c_2 e^{-2x} \\ 2c_1 e^{3x} + c_2 e^{-2x} \end{bmatrix}$$

and in Example 3.3 that the original system has a particular solution

$$Y_i = \begin{bmatrix} -25e^x + 105e^{2x} \\ -10e^x + 45e^{2x} \end{bmatrix}.$$

Thus, our original system has general solution

$$Y = Y_H + Y_i = \begin{bmatrix} 7c_1 e^{3x} + c_2 e^{-2x} - 25e^x + 105e^{2x} \\ 2c_1 e^{3x} + c_2 e^{-2x} - 10e^x + 45e^{2x} \end{bmatrix}.$$

We apply the initial condition to obtain the linear system

$$Y(0) = \begin{bmatrix} 7c_1 + c_2 + 80 \\ 2c_1 + c_2 + 35 \end{bmatrix} = \begin{bmatrix} 7 \\ 17 \end{bmatrix}$$

with solution $c_1 = -11$, $c_2 = 4$. Substituting, we find that our initial value problem has solution

$$Y = \begin{bmatrix} -77e^{3x} + 4e^{-2x} - 25e^x + 105e^{2x} \\ -22e^{3x} + 4e^{-2x} - 10e^x + 45e^{2x} \end{bmatrix}.$$

EXERCISES FOR SECTION 2.3

In each exercise, find a particular solution Y_i of the system $Y' = AY + G(x)$, where A is the matrix of the correspondingly numbered exercise for Section 2.1, and $G(x)$ is as given.

1. $G(x) = \begin{bmatrix} 2e^{8x} \\ 3e^{4x} \end{bmatrix}$.

2. $G(x) = \begin{bmatrix} 2e^{-7x} \\ 6e^{-8x} \end{bmatrix}$.

3. $G(x) = \begin{bmatrix} e^{4x} \\ 4e^{5x} \end{bmatrix}$.

4. $G(x) = \begin{bmatrix} e^{6x} \\ 9e^{8x} \end{bmatrix}$.

5. $G(x) = \begin{bmatrix} 9e^{10x} \\ 25e^{12x} \end{bmatrix}$.

6. $G(x) = \begin{bmatrix} 5e^{-x} \\ 12e^{2x} \end{bmatrix}$.

7. $G(x) = \begin{bmatrix} 1 \\ 3e^{2x} \\ 5e^{4x} \end{bmatrix}$.

8. $G(x) = \begin{bmatrix} 8 \\ 3e^{3x} \\ 3e^{5x} \end{bmatrix}$.

2.4   THE MATRIX EXPONENTIAL

In this section, we will discuss the matrix exponential and its use in solving systems $Y' = AY$.

Our first task is to ask what it means to take a matrix exponential. To answer this, we are guided by ordinary exponentials. Recall that, for any complex number z, the exponential e^z is given by

$$e^z = 1 + z + z^2/2! + z^3/3! + z^4/4! + \dots .$$

With this in mind, we define the matrix exponential as follows.

Definition 4.1. Let T be a square matrix. Then the *matrix exponential* e^T is defined by

$$e^T = I + T + \frac{1}{2!}T^2 + \frac{1}{3!}T^3 + \frac{1}{4!}T^4 + \ldots \quad .$$

(For this definition to make sense we need to know that this series always converges, and it does.)

Recall that the differential equation $y' = ay$ has the solution $y = ce^{ax}$. The situation for $Y' = AY$ is very analogous. (Note that we use Γ rather than C to denote a vector of constants for reasons that will become clear a little later. Note that Γ is on the right in Theorem 4.2 below, a consequence of the fact that matrix multiplication is not commutative.)

Theorem 4.2.

(1) Let A be a square matrix. Then the general solution of

$$Y' = AY$$

is given by

$$Y = e^{Ax}\Gamma$$

where Γ is a vector of arbitrary constants.

(2) The initial value problem

$$Y' = AY, \qquad Y(0) = Y_0$$

has solution

$$Y = e^{Ax}Y_0 \ .$$

Proof. (Outline) (1) We first compute e^{Ax}. In order to do so, note that $(Ax)^2 = (Ax)(Ax) = (AA)(xx) = A^2x^2$ as matrix multiplication commutes with scalar multiplication, and $(Ax)^3 = (Ax)^2(Ax) = (A^2x^2)(Ax) = (A^2A)(x^2x) = A^3x^3$, and similarly, $(Ax)^k = A^kx^k$ for any k. Then, substituting in Definition 4.1, we have that

$$Y = e^{Ax}\Gamma = (I + Ax + \frac{1}{2!}A^2x^2 + \frac{1}{3!}A^3x^3 + \frac{1}{4!}A^4x^4 + \ldots)\Gamma \ .$$

To find Y', we may differentiate this series term-by-term. (This claim requires proof, but we shall not give it here.) Remembering that A and Γ are constant matrices, we see that

$$Y' = (A + \frac{1}{2!}A^2(2x) + \frac{1}{3!}A^3(3x^2) + \frac{1}{4!}A^4(4x^3) + \ldots)\Gamma$$

$$= (A + A^2x + \frac{1}{2!}A^3x^2 + \frac{1}{3!}A^4x^3 + \ldots)\Gamma$$

$$= A(I + Ax + \frac{1}{2!}A^2x^2 + \frac{1}{3!}A^3x^3 + \ldots)\Gamma$$

$$= A(e^{Ax}\Gamma) = AY$$

as claimed.

(2) By (1) we know that $Y' = AY$ has solution $Y = e^{Ax}\Gamma$. We use the initial condition to solve for Γ. Setting $x = 0$, we have:

$$Y_0 = Y(0) = e^{A0}\Gamma = e^0\Gamma = I\Gamma = \Gamma$$

(where e^0 means the exponential of the zero matrix, and the value of this is the identity matrix I, as is apparent from Definition 4.1), so $\Gamma = Y_0$ and $Y = e^{Ax}\Gamma = e^{Ax}Y_0$. $\square$

In the remainder of this section we shall see how to translate the theoretical solution of $Y' = AY$ given by Theorem 4.2 into a practical one. To keep our notation simple, we will stick to 2-by-2 or 3-by-3 cases, but the principle is the same regardless of the size of the matrix.

One case is relatively easy.

Lemma 4.3.   *If J is a diagonal matrix,*

$$J = \begin{bmatrix} d_1 & & & \\ & d_2 & 0 & \\ & 0 & \ddots & \\ & & & d_n \end{bmatrix}$$

then e^{Jx} is the diagonal matrix

$$e^{Jx} = \begin{bmatrix} e^{d_1x} & & & \\ & e^{d_2x} & 0 & \\ & 0 & \ddots & \\ & & & e^{d_nx} \end{bmatrix}.$$

Proof. Suppose, for simplicity, that J is 2-by-2,

$$J = \begin{bmatrix} d_1 & 0 \\ 0 & d_2 \end{bmatrix}.$$

Then you can easily compute that $J^2 = \begin{bmatrix} d_1{}^2 & 0 \\ 0 & d_2{}^2 \end{bmatrix}$, $J^3 = \begin{bmatrix} d_1{}^3 & 0 \\ 0 & d_2{}^3 \end{bmatrix}$, and similarly, $J^k = \begin{bmatrix} d_1{}^k & 0 \\ 0 & d_2{}^k \end{bmatrix}$ for any k.

Then, as in the proof of Theorem 4.2,

$$e^{Jx} = I + Jx + \frac{1}{2!}J^2x^2 + \frac{1}{3!}J^3x^3 + \frac{1}{4!}J^4x^4 + \cdots$$

$$= \begin{bmatrix} 1 & 0 \\ 0 & 1 \end{bmatrix} + \begin{bmatrix} d_1 & 0 \\ 0 & d_2 \end{bmatrix}x + \frac{1}{2!}\begin{bmatrix} d_1{}^2 & 0 \\ 0 & d_2{}^2 \end{bmatrix}x^2 + \frac{1}{3!}\begin{bmatrix} d_1{}^3 & 0 \\ 0 & d_2{}^3 \end{bmatrix}x^3 + \cdots$$

$$= \begin{bmatrix} 1 + d_1x + \frac{1}{2!}(d_1x)^2 + \frac{1}{3!}(d_1x)^3 + \cdots & 0 \\ 0 & 1 + d_2x + \frac{1}{2!}(d_2x)^2 + \frac{1}{3!}(d_2x)^3 + \cdots \end{bmatrix}$$

which we recognize as

$$= \begin{bmatrix} e^{d_1x} & 0 \\ 0 & e^{d_2x} \end{bmatrix}.$$

$\square$

Example 4.4. We wish to find the general solution of $Y' = JY$ where

$$J = \begin{bmatrix} 3 & 0 \\ 0 & -2 \end{bmatrix}.$$

To do so we directly apply Theorem 4.2 and Lemma 4.3. The solution is given by

$$\begin{bmatrix} y_1 \\ y_2 \end{bmatrix} = Y = e^{Jx}\Gamma = \begin{bmatrix} e^{3x} & 0 \\ 0 & e^{-2x} \end{bmatrix}\begin{bmatrix} \gamma_1 \\ \gamma_2 \end{bmatrix} = \begin{bmatrix} \gamma_1 e^{3x} \\ \gamma_2 e^{-2x} \end{bmatrix}.$$

Now suppose we want to find the general solution of $Y' = AY$ where $A = \begin{bmatrix} 5 & -7 \\ 2 & -4 \end{bmatrix}$. We may still apply Theorem 4.2 to conclude that the solution is $Y = e^{Ax}\Gamma$. We again try to calculate e^{Ax}. Now we find

$$A = \begin{bmatrix} 5 & -7 \\ 2 & -4 \end{bmatrix}, \quad A^2 = \begin{bmatrix} 11 & -7 \\ 2 & 2 \end{bmatrix}, \quad A^3 = \begin{bmatrix} 41 & -49 \\ 14 & -22 \end{bmatrix}, \quad \cdots$$

so

$$e^{Ax} = \begin{bmatrix} 1 & 0 \\ 0 & 1 \end{bmatrix} + \begin{bmatrix} 5 & -7 \\ 2 & -4 \end{bmatrix} x + \frac{1}{2!} \begin{bmatrix} 11 & -7 \\ 2 & 2 \end{bmatrix} x^2 + \frac{1}{3!} \begin{bmatrix} 41 & -49 \\ 14 & -22 \end{bmatrix} x^3 + \dots ,$$

which looks like a hopeless mess. But, in fact, the situation is not so hard!

Lemma 4.5.   *Let S and T be two matrices and suppose*

$$S = PTP^{-1}$$

for some invertible matrix P. Then

$$S^k = PT^k P^{-1} \text{ for every } k$$

and

$$e^S = Pe^T P^{-1} .$$

Proof. We simply compute

$$S^2 = SS = (PTP^{-1})(PTP^{-1}) = PT(P^{-1}P)TP^{-1} = PTITP^{-1}$$
$$= PTTP^{-1} = PT^2 P^{-1},$$
$$S^3 = S^2 S = (PT^2 P^{-1})(PTP^{-1}) = PT^2(P^{-1}P)TP^{-1} = PT^2 ITP^{-1}$$
$$= PT^2 TP^{-1} = PT^3 P^{-1},$$
$$S^4 = S^3 S = (PT^3 P^{-1})(PTP^{-1}) = PT^3(P^{-1}P)TP^{-1} = PT^3 ITP^{-1}$$
$$= PT^3 TP^{-1} = PT^4 P^{-1} ,$$

etc.

Then

$$e^S = I + S + \frac{1}{2!} S^2 + \frac{1}{3!} S^3 + \frac{1}{4!} S^4 + \dots$$
$$= PIP^{-1} + PTP^{-1} + \frac{1}{2!} PT^2 P^{-1} + \frac{1}{3!} PT^3 P^{-1} + \frac{1}{4!} PT^4 P^{-1} + \dots$$
$$= P(I + T + \frac{1}{2!} T^2 + \frac{1}{3!} T^3 + \frac{1}{4!} T^4 + \dots)P^{-1}$$
$$= Pe^T P^{-1}$$

as claimed. $\square$

With this in hand let us return to our problem.

Example 4.6.   (Compare Example 1.2.) We wish to find the general solution of $Y' = AY$ where

$$A = \begin{bmatrix} 5 & -7 \\ 2 & -4 \end{bmatrix}.$$

We saw in Example 1.16 in Chapter 1 that $A = PJP^{-1}$ with

$$P = \begin{bmatrix} 7 & 1 \\ 2 & 1 \end{bmatrix} \text{ and } J = \begin{bmatrix} 3 & 0 \\ 0 & -2 \end{bmatrix}.$$

Then

$$\begin{aligned}
e^{Ax} &= Pe^{Jx}P^{-1} \\
&= \begin{bmatrix} 7 & 1 \\ 2 & 1 \end{bmatrix} \begin{bmatrix} e^{3x} & 0 \\ 0 & e^{-2x} \end{bmatrix} \begin{bmatrix} 7 & 1 \\ 2 & 1 \end{bmatrix}^{-1} \\
&= \begin{bmatrix} \frac{7}{5}e^{3x} - \frac{2}{5}e^{-2x} & -\frac{7}{5}e^{3x} + \frac{7}{5}e^{-2x} \\ \frac{2}{5}e^{3x} - \frac{2}{5}e^{-2x} & -\frac{2}{5}e^{3x} + \frac{7}{5}e^{-2x} \end{bmatrix}
\end{aligned}$$

and

$$\begin{aligned}
Y = e^{Ax}\Gamma = e^{Ax} \begin{bmatrix} \gamma_1 \\ \gamma_2 \end{bmatrix} \\
= \begin{bmatrix} (\frac{7}{5}\gamma_1 - \frac{7}{5}\gamma_2)e^{3x} + (-\frac{2}{5}\gamma_1 + \frac{7}{5}\gamma_2)e^{-2x} \\ (\frac{2}{5}\gamma_1 - \frac{2}{5}\gamma_2)e^{3x} + (-\frac{2}{5}\gamma_1 + \frac{7}{5}\gamma_2)e^{-2x} \end{bmatrix}.
\end{aligned}$$

Example 4.7.   (Compare Example 1.3.) We wish to find the general solution of $Y' = AY$ where

$$A = \begin{bmatrix} 2 & -3 & -3 \\ 2 & -2 & -2 \\ -2 & 1 & 1 \end{bmatrix}.$$

We saw in Example 2.23 in Chapter 1 that $A = PJP^{-1}$ with

$$P = \begin{bmatrix} 1 & 0 & -1 \\ 0 & -1 & -1 \\ 1 & 1 & 1 \end{bmatrix} \text{ and } J = \begin{bmatrix} -1 & 0 & 0 \\ 0 & 0 & 0 \\ 0 & 0 & 2 \end{bmatrix}.$$

Then

$$e^{Ax} = Pe^{Jx}P^{-1}$$

$$= \begin{bmatrix} 1 & 0 & -1 \\ 0 & -1 & -1 \\ 1 & 1 & 1 \end{bmatrix} \begin{bmatrix} e^{-x} & 0 & 0 \\ 0 & 1 & 0 \\ 0 & 0 & e^{2x} \end{bmatrix} \begin{bmatrix} 1 & 0 & -1 \\ 0 & -1 & -1 \\ 1 & 1 & 1 \end{bmatrix}^{-1}$$

$$= \begin{bmatrix} e^{2x} & e^{-x} - e^{2x} & e^{-x} - e^{2x} \\ -1 + e^{2x} & 2 - e^{2x} & 1 - e^{2x} \\ 1 - e^{2x} & e^{-x} - 2 + e^{2x} & e^{-x} - 1 + e^{2x} \end{bmatrix}$$

and

$$Y = e^{Ax}\Gamma = e^{Ax} \begin{bmatrix} \gamma_1 \\ \gamma_2 \\ \gamma_3 \end{bmatrix}$$

$$= \begin{bmatrix} (\gamma_2 + \gamma_3)e^{-x} + (\gamma_1 - \gamma_2 - \gamma_3)e^{2x} \\ (-\gamma_1 + 2\gamma_2 + \gamma_3) + (\gamma_1 - \gamma_2 - \gamma_3)e^{2x} \\ (\gamma_2 + \gamma_3)e^{-x} + (\gamma_1 - 2\gamma_2 - \gamma_3) + (-\gamma_1 + \gamma_2 + \gamma_3)e^{2x} \end{bmatrix}.$$

Now suppose we want to solve the initial value problem $Y' = AY, Y(0) = \begin{bmatrix} 1 \\ 0 \\ 0 \end{bmatrix}$. Then

$$Y = e^{Ax}Y(0)$$

$$= \begin{bmatrix} e^{2x} & e^{-x} - e^{2x} & e^{-x} - e^{2x} \\ -1 + e^{2x} & 2 - e^{2x} & 1 - e^{2x} \\ 1 - e^{2x} & e^{-x} - 2 + e^{2x} & e^{-x} - 1 + e^{2x} \end{bmatrix} \begin{bmatrix} 1 \\ 0 \\ 0 \end{bmatrix}$$

$$= \begin{bmatrix} e^{2x} \\ -1 + e^{2x} \\ 1 - e^{2x} \end{bmatrix}.$$

Remark 4.8. Let us compare the results of our method here with that of our previous method. In the case of Example 4.6, our previous method gives the solution

$$Y = P \begin{bmatrix} e^{3x} & 0 \\ 0 & e^{-2x} \end{bmatrix} C$$

$$= Pe^{Jx}C$$

where $J = \begin{bmatrix} 3 & 0 \\ 0 & -2 \end{bmatrix}$,

while our method here gives

$$Y = P e^{Jx} P^{-1} \Gamma \,.$$

But note that these answers are really the same! For P^{-1} is a constant matrix, so if Γ is a vector of arbitrary constants, then so is $P^{-1}\Gamma$, and we simply set $C = P^{-1}\Gamma$.

Similarly, in the case of Example 4.7, our previous method gives the solution

$$Y = P \begin{bmatrix} e^{-x} & 0 & 0 \\ 0 & 1 & 0 \\ 0 & 0 & e^{2x} \end{bmatrix} C$$

$$= P e^{Jx} C$$

where $J = \begin{bmatrix} -1 & 0 & 0 \\ 0 & 0 & 0 \\ 0 & 0 & 2 \end{bmatrix}$,

while our method here gives

$$Y = P e^{Jx} P^{-1} \Gamma$$

and again, setting $C = P^{-1}\Gamma$, we see that these answers are the same.

So the point here is not that the matrix exponential enables us to solve new problems, but rather that it gives a new viewpoint about the solutions that we have already obtained.

While these two methods are in principle the same, we may ask which is preferable in practice. In this regard we see that our earlier method is better, as the use of the matrix exponential requires us to find P^{-1}, which may be a considerable amount of work. However, this advantage is (partially) negated if we wish to solve initial value problems, as the matrix exponential method immediately gives the unknown constants Γ, as $\Gamma = Y(0)$, while in the former method we must solve a linear system to obtain the unknown constants C.

Now let us consider the nondiagonalizable case. Suppose $Z' = JZ$ where J is a matrix consisting of a single Jordan block. Then by Theorem 4.2 this has the solution $Z = e^{Jx}\Gamma$. On the other hand, in Theorem 1.1 we already saw that this system has solution $Z = M_Z C$. In this case, we simply have $C = \Gamma$, so we must have $e^{Jx} = M_Z$. Let us see that this is true by computing e^{Jx} directly.

Theorem 4.9.   *Let J be a k–by–k Jordan block with eigenvalue a,*

$$
J = \begin{bmatrix}
a & 1 & & & & \\
 & a & 1 & & 0 & \\
 & & a & 1 & & \\
 & & & \ddots & \ddots & \\
 & 0 & & & a & 1 \\
 & & & & & a
\end{bmatrix}.
$$

Then

$$
e^{Jx} = e^{ax} \begin{bmatrix}
1 & x & x^2/2! & x^3/3! & \cdots & x^{k-1}/(k-1)! \\
0 & 1 & x & x^2/2! & \cdots & x^{k-2}/(k-2)! \\
 & & 1 & x & \cdots & x^{k-3}/(k-3)! \\
 & & & \ddots & & \vdots \\
 & & & & & x \\
 & & & & & 1
\end{bmatrix}.
$$

Proof. First suppose that J is a 2-by-2 Jordan block,

$$
J = \begin{bmatrix} a & 1 \\ 0 & a \end{bmatrix}.
$$

Then $J^2 = \begin{bmatrix} a^2 & 2a \\ 0 & a^2 \end{bmatrix}$, $J^3 = \begin{bmatrix} a^3 & 3a^2 \\ 0 & a^3 \end{bmatrix}$, $J^4 = \begin{bmatrix} a^4 & 4a^3 \\ 0 & a^4 \end{bmatrix}, \ldots$ so

$$
e^{Jx} = \begin{bmatrix} 1 & 0 \\ 0 & 1 \end{bmatrix} + \begin{bmatrix} a & 1 \\ 0 & a \end{bmatrix} x + \frac{1}{2!} \begin{bmatrix} a^2 & 2a \\ 0 & a^2 \end{bmatrix} x^2 + \frac{1}{3!} \begin{bmatrix} a^3 & 3a^2 \\ 0 & a^3 \end{bmatrix} x^3 + \frac{1}{4!} \begin{bmatrix} a^4 & 4a^3 \\ 0 & a^4 \end{bmatrix} x^4 + \ldots
$$

$$
= \begin{bmatrix} m_{11} & m_{12} \\ 0 & m_{22} \end{bmatrix},
$$

and we see that

$$
m_{11} = m_{22} = 1 + ax + \frac{1}{2!}(ax)^2 + \frac{1}{3!}(ax)^3 + \frac{1}{4!}(ax)^4 + \frac{1}{5!}(ax)^5 + \ldots
$$

$$
= e^{ax},
$$

and

$$m_{12} = x + ax^2 + \frac{1}{2!}a^2x^3 + \frac{1}{3!}a^3x^4 + \frac{1}{4!}a^4x^5 + \ldots$$

$$= x(1 + ax + \frac{1}{2!}(ax)^2 + \frac{1}{3!}(ax)^3 + \frac{1}{4!}(ax)^4 + \ldots) = xe^{ax}$$

and so we conclude that

$$e^{Jx} = \begin{bmatrix} e^{ax} & xe^{ax} \\ 0 & e^{ax} \end{bmatrix} = e^{ax}\begin{bmatrix} 1 & x \\ 0 & 1 \end{bmatrix}.$$

Next suppose that J is a 3-by-3 Jordan block,

$$J = \begin{bmatrix} a & 1 & 0 \\ 0 & a & 1 \\ 0 & 0 & 1 \end{bmatrix}.$$

Then $J^2 = \begin{bmatrix} a^2 & 2a & 1 \\ 0 & a^2 & 2a \\ 0 & 0 & a^2 \end{bmatrix}$, $J^3 = \begin{bmatrix} a^3 & 3a^2 & 3a \\ 0 & a^3 & 3a^2 \\ 0 & 0 & a^3 \end{bmatrix}$, $J^4 = \begin{bmatrix} a^4 & 4a^3 & 6a^2 \\ 0 & a^4 & 4a^3 \\ 0 & 0 & a^4 \end{bmatrix}$,

$$J^5 = \begin{bmatrix} a^5 & 5a^4 & 10a^3 \\ 0 & a^5 & 5a^4 \\ 0 & 0 & a^5 \end{bmatrix}, \ldots$$

so

$$e^{Jx} = \begin{bmatrix} 1 & 0 & 0 \\ 0 & 1 & 0 \\ 0 & 0 & 1 \end{bmatrix} + \begin{bmatrix} a & 1 & 0 \\ 0 & a & 1 \\ 0 & 0 & a \end{bmatrix}x + \frac{1}{2!}\begin{bmatrix} a^2 & 2a & 1 \\ 0 & a^2 & 2a \\ 0 & 0 & a^2 \end{bmatrix}x^2 + \frac{1}{3!}\begin{bmatrix} a^3 & 3a^2 & 3a \\ 0 & a^3 & 3a^2 \\ 0 & 0 & a^3 \end{bmatrix}x^3$$

$$+ \frac{1}{4!}\begin{bmatrix} a^4 & 4a^3 & 6a^2 \\ 0 & a^4 & 4a^3 \\ 0 & 0 & a^4 \end{bmatrix}x^4 + \frac{1}{5!}\begin{bmatrix} a^5 & 5a^4 & 10a^3 \\ 0 & a^5 & 5a^4 \\ 0 & 0 & a^5 \end{bmatrix}x^5 + \ldots$$

$$= \begin{bmatrix} m_{11} & m_{12} & m_{13} \\ 0 & m_{22} & m_{23} \\ 0 & 0 & m_{33} \end{bmatrix},$$

and we see that

$$m_{11} = m_{22} = m_{33} = 1 + ax + \frac{1}{2!}(ax)^2 + \frac{1}{3!}(ax)^3 + \frac{1}{4!}(ax)^4 + \frac{1}{5!}(ax)^5 + \ldots$$

$$= e^{ax},$$

and

$$m_{12} = m_{23} = x + ax^2 + \frac{1}{2!}a^2x^3 + \frac{1}{3!}a^3x^4 + \frac{1}{4!}a^4x^5 + \ldots$$

$$= xe^{ax}$$

as we saw in the 2-by-2 case. Finally,

$$m_{13} = \frac{1}{2!}x^2 + \frac{1}{2!}ax^3 + \frac{1}{2!}(\frac{1}{2!}a^2x^4) + \frac{1}{2!}(\frac{1}{3!}a^3x^5) + \ldots$$

(as $6/4! = 1/4 = (1/2!)(1/2!)$ and $10/5! = 1/12 = (1/2!)(1/3!)$, etc.)

$$= \frac{1}{2!}x^2(1 + ax + \frac{1}{2!}(ax)^2 + \frac{1}{3!}(ax)^3 + \ldots)$$

$$= \frac{1}{2!}x^2e^{ax} ,$$

so

$$e^{Jx} = \begin{bmatrix} e^{ax} & xe^{ax} & \frac{1}{2!}x^2e^{ax} \\ 0 & e^{ax} & xe^{ax} \\ 0 & 0 & e^{ax} \end{bmatrix} = e^{ax}\begin{bmatrix} 1 & x & x^2/2! \\ 0 & 1 & x \\ 0 & 0 & 1 \end{bmatrix} ,$$

and similarly, for larger Jordan blocks.   □

Let us see how to apply this theorem in a couple of examples.

Example 4.10.   (Compare Examples 1.6 and 1.10.) Consider the system

$$Y' = AY \text{ where } A = \begin{bmatrix} 0 & 1 \\ -4 & 4 \end{bmatrix} .$$

Also, consider the initial value problem $Y' = AY, Y(0) = \begin{bmatrix} 3 \\ -8 \end{bmatrix}$.

We saw in Example 2.12 in Chapter 1 that $A = PJP^{-1}$ with

$$P = \begin{bmatrix} -2 & 1 \\ -4 & 0 \end{bmatrix} \text{ and } J = \begin{bmatrix} 2 & 1 \\ 0 & 2 \end{bmatrix} .$$

Then

$$e^{Ax} = Pe^{Jx}P^{-1}$$

$$= \begin{bmatrix} -2 & 1 \\ -4 & 0 \end{bmatrix}\begin{bmatrix} e^{2x} & xe^{2x} \\ 0 & e^{2x} \end{bmatrix}\begin{bmatrix} -2 & 1 \\ -4 & 0 \end{bmatrix}^{-1}$$

$$= \begin{bmatrix} (1-2x)e^{2x} & xe^{2x} \\ -4xe^{2x} & (1+2x)e^{2x} \end{bmatrix} ,$$

and so

$$Y = e^{Ax}\Gamma$$

$$= e^{Ax}\begin{bmatrix} \gamma_1 \\ \gamma_2 \end{bmatrix}$$

$$= \begin{bmatrix} \gamma_1 e^{2x} + (-2\gamma_1 + \gamma_2)xe^{2x} \\ \gamma_2 e^{2x} + (-4\gamma_1 + 2\gamma_2)xe^{2x} \end{bmatrix}.$$

The initial value problem has solution

$$Y = e^{Ax}Y_0$$

$$= e^{Ax}\begin{bmatrix} 3 \\ -8 \end{bmatrix}$$

$$= \begin{bmatrix} 3e^{2x} - 14xe^{2x} \\ 8e^{2x} - 28xe^{2x} \end{bmatrix}.$$

Example 4.11.   (Compare Examples 1.7 and 1.11.) Consider the system

$$Y' = AY \text{ where } A = \begin{bmatrix} 2 & 1 & 1 \\ 2 & 1 & -2 \\ -1 & 0 & 2 \end{bmatrix}.$$

Also, consider the initial value problem $Y' = AY, Y(0) = \begin{bmatrix} 8 \\ 32 \\ 5 \end{bmatrix}.$

We saw in Example 2.25 in Chapter 1 that $A = PJP^{-1}$ with

$$P = \begin{bmatrix} 1 & 0 & -5 \\ -2 & 0 & 6 \\ -1 & 1 & 1 \end{bmatrix} \text{ and } J = \begin{bmatrix} -1 & 1 & 0 \\ 0 & -1 & 0 \\ 0 & 0 & 3 \end{bmatrix}.$$

Then

$$e^{Ax} = Pe^{Jx}P^{-1}$$

$$= \begin{bmatrix} 1 & 0 & -5 \\ -2 & 0 & 6 \\ -1 & 1 & 1 \end{bmatrix}\begin{bmatrix} e^{-x} & xe^{-x} & 0 \\ 0 & e^{-x} & 0 \\ 0 & 0 & e^{3x} \end{bmatrix}\begin{bmatrix} 1 & 0 & -5 \\ -2 & 0 & 6 \\ -1 & 1 & 1 \end{bmatrix}^{-1}$$

$$= \begin{bmatrix} \frac{3}{8}e^{-x} + \frac{1}{2}xe^{-x} + \frac{5}{8}e^{3x} & -\frac{5}{16}e^{-x} - \frac{1}{4}xe^{-x} + \frac{5}{16}e^{3x} & xe^{-x} \\ -\frac{3}{4}e^{-x} - xe^{-x} + \frac{3}{4}e^{3x} & \frac{5}{8}e^{-x} + \frac{1}{2}xe^{-x} + \frac{3}{8}e^{3x} & -2xe^{-x} \\ \frac{1}{8}e^{-x} - \frac{1}{2}xe^{-x} - \frac{1}{8}e^{3x} & \frac{1}{16}e^{-x} + \frac{1}{4}xe^{-x} - \frac{1}{16}e^{3x} & e^{-x} - xe^{-x} \end{bmatrix}$$

and so

$$Y = e^{Ax}\Gamma$$

$$= e^{Ax}\begin{bmatrix} \gamma_1 \\ \gamma_2 \\ \gamma_3 \end{bmatrix}$$

$$= \begin{bmatrix} (\frac{3}{8}\gamma_1 - \frac{5}{16}\gamma_2)e^{-x} + (\frac{1}{2}\gamma_1 - \frac{1}{4}\gamma_2 + \gamma_3)xe^{-x} + (\frac{5}{8}\gamma_1 + \frac{5}{16}\gamma_2)e^{3x} \\ (-\frac{3}{4}\gamma_1 + \frac{5}{8}\gamma_2)e^{-x} + (-\gamma_1 + \frac{1}{2}\gamma_2 - 2\gamma_3)xe^{-x} + (\frac{3}{4}\gamma_1 + \frac{3}{8}\gamma_2)e^{3x} \\ (\frac{1}{8}\gamma_1 + \frac{1}{16}\gamma_2 + \gamma_3)e^{-x} + (-\frac{1}{2}\gamma_1 + \frac{1}{4}\gamma_2 - \gamma_3)xe^{-x} + (-\frac{1}{8}\gamma_1 - \frac{1}{16}\gamma_2)e^{3x} \end{bmatrix}.$$

The initial value problem has solution

$$Y = e^{Ax}Y_0$$

$$= e^{Ax}\begin{bmatrix} 8 \\ 32 \\ 5 \end{bmatrix}$$

$$= \begin{bmatrix} -7e^{-x} + xe^{-x} + 15e^{3x} \\ 14e^{-x} - 2xe^{-x} + 18e^{3x} \\ 8e^{-x} - xe^{-x} - 3e^{3x} \end{bmatrix}.$$

Now we solve $Y' = AY$ in an example where the matrix A has complex eigenvalues. As you will see, our method is exactly the same.

Example 4.12. (Compare Example 2.4.) Consider the system

$$Y' = AY \text{ where } A = \begin{bmatrix} 2 & -17 \\ 1 & 4 \end{bmatrix}.$$

We saw in Example 2.4 that $A = PJP^{-1}$ with

$$P = \begin{bmatrix} -1 + 4i & -1 - 4i \\ 1 & 1 \end{bmatrix} \text{ and } J = \begin{bmatrix} 3 + 4i & 0 \\ 0 & 3 - 4i \end{bmatrix}.$$

Then

$$e^{Ax} = Pe^{Jx}P^{-1}$$

$$= \begin{bmatrix} -1+4i & -1-4i \\ 1 & 1 \end{bmatrix} \begin{bmatrix} e^{(3+4i)x} & 0 \\ 0 & e^{(3-4i)x} \end{bmatrix} \begin{bmatrix} -1+4i & -1-4i \\ 1 & 1 \end{bmatrix}^{-1}$$

$$= \begin{bmatrix} -1+4i & -1-4i \\ 1 & 1 \end{bmatrix} \begin{bmatrix} e^{(3+4i)x} & 0 \\ 0 & e^{(3-4i)x} \end{bmatrix} (1/(8i)) \begin{bmatrix} 1 & 1+4i \\ -1 & -1+4i \end{bmatrix}$$

$$= \begin{bmatrix} m_{11} & m_{12} \\ m_{21} & m_{22} \end{bmatrix}$$

where

$$
\begin{aligned}
m_{11} &= (1/(8i))((-1+4i)e^{(3+4i)x} + (-1-4i)(-e^{(3-4i)x})) \\
&= (1/(8i))((-1+4i)e^{3x}(\cos(4x)+i\sin(4x)) - (-1-4i)e^{3x}(\cos(4x)-i\sin(4x)) \\
&= (1/(8i))(ie^{3x}(4\cos(4x)-\sin(4x))(2)) = e^{3x}(\cos(4x)-(1/4)\sin(4x)), \\
m_{12} &= (1/(8i))((-1+4i)(1+4i)e^{(3+4i)x} + (-1-4i)(-1+4i)e^{(3-4i)x}) \\
&= (1/(8i))(ie^{3x}(-17\sin(4x))(2)) = e^{3x}((-17/4)\sin(4x)), \\
m_{21} &= (1/(8i))(e^{(3+4i)x} - e^{(3-4i)x}) \\
&= (1/(8i))(ie^{3x}(\sin(4x))(2)) = e^{3x}((1/4)\sin(4x)), \\
m_{22} &= (1/(8i))((1+4i)e^{(3+4i)x} + (-1+4i)e^{(3-4i)x}) \\
&= (1/(8i))((1+4i)e^{3x}(\cos(4x)+i\sin(4x)) + (-1+4i)e^{3x}(\cos(4x)-i\sin(4x)) \\
&= (1/(8i))(ie^{3x}(4\cos(4x)+\sin(4x))(2)) = e^{3x}(\cos(4x)+(1/4)\sin(4x)) .
\end{aligned}
$$

Thus,

$$e^{Ax} = e^{3x} \begin{bmatrix} \cos(4x) - \frac{1}{4}\sin(4x) & \frac{-17}{4}\sin(4x) \\ \frac{1}{4}\sin(4x) & \cos(4x) + \frac{1}{4}\sin(4x) \end{bmatrix}$$

and

$$Y = e^{Ax}\Gamma = \begin{bmatrix} \gamma_1 e^{3x}\cos(4x) + (\frac{-1}{4}\gamma_1 + \frac{-17}{4}\gamma_2)e^{3x}\sin(4x) \\ \gamma_2 e^{3x}\cos(4x) + (\frac{1}{4}\gamma_1 + \frac{1}{4}\gamma_2)e^{3x}\sin(4x) \end{bmatrix} .$$

Remark 4.13. Our procedure in this section is essentially that of Remark 1.12. (Compare Example 4.10 with Example 1.13.)

Remark 4.14. As we have seen, for a matrix J in JCF, $e^{Jx} = M_Z$, in the notation of Section 2.1. But also, in the notation of Section 2.1, if $A = PJP^{-1}$, then $e^{Ax} = Pe^{Jx}P^{-1} = PM_ZP^{-1} = M_Y$.

Remark 4.15.   Now let us see how to use the matrix exponential to solve an inhomogeneous system $Y' = AY + G(x)$. Since we already know how to solve homogeneous systems, we need only, by Lemma 3.1, find a (single) particular solution Y_i of this inhomogeneous system, and that is what we do. We shall again use our notation from Section 2.3, that $\int_0 H(x)dx$ denotes an arbitrary (but fixed) antiderivative of $H(x)$.

Thus, consider $Y' = AY + G(x)$. Then, proceeding analogously as for an ordinary first-order linear differential equation, we have

$$Y' = AY + G(x)$$
$$Y' - AY = G(x)$$

and, multiplying this equation by the integrating factor e^{-Ax}, we obtain

$$e^{-Ax}(Y' - AY) = e^{-Ax}G(x)$$
$$(e^{-Ax}Y)' = e^{-Ax}G(x)$$

with solution

$$e^{-Ax}Y_i = \int_0 e^{-Ax}G(x)$$
$$Y_i = e^{Ax}\int_0 e^{-Ax}G(x) \, .$$

Let us compare this with the solution we found in Theorem 3.2. By Remark 4.14, we can rewrite this solution as $Y_i = M_Y \int_0 M_Y^{-1}G(x)$. This is almost, but not quite, what we had in Theorem 3.2. There we had the solution $Y_i = N_Y \int_0 N_Y^{-1}G(x)$, where $N_Y = PM_Z$. But these solutions are the same, as $M_Y = PM_Z P^{-1} = N_Y P^{-1}$. Then $M_Y^{-1} = PM_Z^{-1}P^{-1}$ and $N_Y^{-1} = M_Z^{-1}P^{-1}$, so $M_Y^{-1} = PN_Y^{-1}$. Substituting, we find

$$Y_i = M_Y \int_0 M_Y^{-1}G(x)$$
$$= N_Y P^{-1} \int_0 PN_Y^{-1}G(x) \, ,$$

and, since P is a constant matrix, we may bring it outside the integral to obtain

$$Y_i = N_Y P^{-1}P \int_0 N_Y^{-1}G(x)$$
$$= N_Y \int_0 N_Y^{-1}G(x)$$

as claimed.

Remark 4.16. In applying this method we must compute $M_Z^{-1} = (e^{Jx})^{-1} = e^{-Jx} = e^{J(-x)}$, and, as an aid to calculation, it is convenient to make the following observation. Suppose, for simplicity, that J consists of a single Jordan block. Then we compute: in the 1-by-1 case, $\left[e^{ax}\right]^{-1} = \left[e^{-ax}\right]$; in the 2-by-2 case, $\left(e^{ax}\begin{bmatrix} 1 & x \\ 0 & 1 \end{bmatrix}\right)^{-1} = e^{-ax}\begin{bmatrix} 1 & -x \\ 0 & 1 \end{bmatrix}$; in the 3-by-3 case,

$$\left(e^{ax}\begin{bmatrix} 1 & x & x^2/2! \\ 0 & 1 & x \\ 0 & 0 & 1 \end{bmatrix}\right)^{-1} = e^{-ax}\begin{bmatrix} 1 & -x & (-x)^2/2! \\ 0 & 1 & -x \\ 0 & 0 & 1 \end{bmatrix} = e^{-ax}\begin{bmatrix} 1 & -x & x^2/2! \\ 0 & 1 & -x \\ 0 & 0 & 1 \end{bmatrix}, \text{etc.}$$

EXERCISES FOR SECTION 2.4

In each exercise:

(a) Find e^{Ax} and the solution $Y = e^{Ax}\Gamma$ of $Y' = AY$.

(b) Use part (a) to solve the initial value problem $Y' = AY, Y(0) = Y_0$.

Exercises 1–24: In Exercise n, for $1 \leq n \leq 20$, the matrix A and the initial vector Y_0 are the same as in Exercise n of Section 2.1. In Exercise n, for $21 \leq n \leq 24$, the matrix A and the initial vector Y_0 are the same as in Exercise $n - 20$ of Section 2.2.

APPENDIX A

Background Results

A.1 BASES, COORDINATES, AND MATRICES

In this section of the Appendix, we review the basic facts on bases for vector spaces and on coordinates for vectors and matrices for linear transformations. Then we use these to (re)prove some of the results in Chapter 1.

First we see how to represent vectors, once we have chosen a basis.

Theorem 1.1. *Let V be a vector space and let $\mathcal{B} = \{v_1, \ldots, v_n\}$ be a basis of V. Then any vector v in V can be written as $v = c_1 v_1 + \ldots + c_n v_n$ in a unique way.*

This theorem leads to the following definition.

Definition 1.2. Let V be a vector space and let $\mathcal{B} = \{v_1, \ldots, v_n\}$ be a basis of V. Let v be a vector in V and write $v = c_1 v_1 + \ldots + c_n v_n$. Then the vector

$$[v]_\mathcal{B} = \begin{bmatrix} c_1 \\ \vdots \\ c_n \end{bmatrix}$$

is the *coordinate vector* of v in the basis $\mathcal{B}$.

Remark 1.3. In particular, we may take $V = \mathbb{C}^n$ and consider the *standard basis* $\mathcal{E} = \{e_1, \ldots, e_n\}$ where

$$e_i = \begin{bmatrix} 0 \\ \vdots \\ 0 \\ 1 \\ 0 \\ \vdots \end{bmatrix},$$

with 1 in the ith position, and 0 elsewhere.

Then, if

$$v = \begin{bmatrix} c_1 \\ c_2 \\ \vdots \\ c_{n-1} \\ c_n \end{bmatrix},$$

we see that

$$v = c_1 \begin{bmatrix} 1 \\ 0 \\ \vdots \\ 0 \\ 0 \end{bmatrix} + \ldots + c_n \begin{bmatrix} 0 \\ 0 \\ \vdots \\ 0 \\ 1 \end{bmatrix} = c_1 e_1 + \ldots + c_n e_n$$

so we then see that

$$[v]_\mathcal{E} = \begin{bmatrix} c_1 \\ c_2 \\ \vdots \\ c_{n-1} \\ c_n \end{bmatrix}.$$

(In other words, a vector in $\mathbb{C}^n$ "looks like" itself in the standard basis.)

Next we see how to represent linear transformations, once we have chosen a basis.

Theorem 1.4.   *Let V be a vector space and let $\mathcal{B} = \{v_1, \ldots, v_n\}$ be a basis of V. Let $T : V \longrightarrow V$ be a linear transformation. Then there is a unique matrix $[T]_\mathcal{B}$ such that, for any vector v in V,*

$$[T(v)]_\mathcal{B} = [T]_\mathcal{B}[v]_\mathcal{B}.$$

Furthermore, the matrix $[T]_\mathcal{B}$ is given by

$$[T]_\mathcal{B} = \left[[v_1]_\mathcal{B} \,\middle|\, [v_2]_\mathcal{B} \,\middle|\, \cdots \,\middle|\, [v_n]_\mathcal{B} \right].$$

Similarly, this theorem leads to the following definition.

Definition 1.5.   Let V be a vector space and let $\mathcal{B} = \{v_1, \ldots, v_n\}$ be a basis of V. Let $T : V \longrightarrow V$ be a linear transformation. Let $[T]_\mathcal{B}$ be the matrix defined in Theorem 1.4. Then $[T]_\mathcal{B}$ is the *matrix of the linear transformation T in the basis $\mathcal{B}$.*

Remark 1.6. In particular, we may take $V = \mathbb{C}^n$ and consider the *standard basis* $\mathcal{E} = \{e_1, \ldots, e_n\}$. Let A be an n-by-n square matrix and write $A = \left[a_1 \,\middle|\, a_2 \,\middle|\, \ldots \,\middle|\, a_n \right]$. If $\mathcal{T}_A$ is the linear transformation given by $\mathcal{T}_A(v) = Av$, then

$$[\mathcal{T}_A]_{\mathcal{E}} = \left[[\mathcal{T}_A(e_1)]_{\mathcal{E}} \,\middle|\, [\mathcal{T}_A(e_2)]_{\mathcal{E}} \,\middle|\, \ldots \,\middle|\, [\mathcal{T}_A(e_n)]_{\mathcal{E}} \right]$$
$$= \left[[Ae_1]_{\mathcal{E}} \,\middle|\, [Ae_2]_{\mathcal{E}} \,\middle|\, \ldots \,\middle|\, [Ae_n]_{\mathcal{E}} \right]$$
$$= \left[[a_1]_{\mathcal{E}} \,\middle|\, [a_2]_{\mathcal{E}} \,\middle|\, \ldots \,\middle|\, [a_n]_{\mathcal{E}} \right]$$
$$= \left[a_1 \,\middle|\, a_2 \,\middle|\, \ldots \,\middle|\, a_n \right]$$
$$= A \, .$$

(In other words, the linear transformation given by multiplication by a matrix "looks like" that same matrix in the standard basis.)

What is essential to us is the ability to compare the situation in different bases. To that end, we have the following theorem.

Theorem 1.7. *Let V be a vector space, and let $\mathcal{B} = \{v_1, \ldots, v_n\}$ and $\mathcal{C} = \{w_1, \ldots, w_n\}$ be two bases of V. Let $P_{\mathcal{C} \leftarrow \mathcal{B}}$ be the matrix*

$$P_{\mathcal{C} \leftarrow \mathcal{B}} = \left[[v_1]_{\mathcal{C}} \,\middle|\, [v_2]_{\mathcal{C}} \,\middle|\, \ldots \,\middle|\, [v_n]_{\mathcal{C}} \right].$$

This matrix has the following properties:

(1) For any vector v in V,

$$[v]_{\mathcal{C}} = P_{\mathcal{C} \leftarrow \mathcal{B}}[v]_{\mathcal{B}} \, .$$

(2) This matrix is invertible and

$$(P_{\mathcal{C} \leftarrow \mathcal{B}})^{-1} = P_{\mathcal{B} \leftarrow \mathcal{C}} = \left[[w_1]_{\mathcal{B}} \,\middle|\, [w_2]_{\mathcal{B}} \,\middle|\, \ldots \,\middle|\, [w_n]_{\mathcal{B}} \right].$$

(3) For any linear transformation $T : V \longrightarrow V$,

$$[T]_{\mathcal{C}} = P_{\mathcal{C} \leftarrow \mathcal{B}}[T]_{\mathcal{B}} P_{\mathcal{B} \leftarrow \mathcal{C}}$$
$$= P_{\mathcal{C} \leftarrow \mathcal{B}}[T]_{\mathcal{B}} (P_{\mathcal{C} \leftarrow \mathcal{B}})^{-1}$$
$$= (P_{\mathcal{B} \leftarrow \mathcal{C}})^{-1}[T]_{\mathcal{B}} P_{\mathcal{B} \leftarrow \mathcal{C}} \, .$$

Again, this theorem leads to a definition.

Definition 1.8.   The matrix $P_{C \leftarrow B}$ is the *change-of-basis matrix* from the basis B to the basis C.

Now we come to what is the crucial point for us.

Corollary 1.9.   *Let $V = \mathbb{C}^n$, let $\mathcal{E}$ be the standard basis of V, and let $B = \{v_1, \ldots, v_n\}$ be any basis of V. Let A be any n-by-n square matrix. Then*

$$A = PBP^{-1}$$

where

$$P = \left[v_1 \,\middle|\, v_2 \,\middle|\, \ldots \,\middle|\, v_n \right] \text{ and } B = [\mathcal{T}_A]_B \,.$$

Proof. By Theorem 1.7,

$$[\mathcal{T}_A]_\mathcal{E} = P_{\mathcal{E} \leftarrow B} [\mathcal{T}_A]_B (P_{\mathcal{E} \leftarrow B})^{-1} \,.$$

But by Remark 1.3,

$$P_{\mathcal{E} \leftarrow B} = \left[[v_1]_\mathcal{E} \,\middle|\, [v_2]_\mathcal{E} \,\middle|\, \ldots \,\middle|\, [v_n]_\mathcal{E} \right] = \left[v_1 \,\middle|\, v_2 \,\middle|\, \ldots \,\middle|\, v_n \right] = P \,,$$

and by Remark 1.6,

$$[\mathcal{T}_A]_\mathcal{E} = A \,.$$

$\square$

With this in hand we now present new proofs of Theorems 1.14 and 2.11 in Chapter 1, and a proof of Lemma 1.7 in Chapter 1. For convenience, we restate these results.

Theorem 1.10.   *Let A be an n-by-n matrix over the complex numbers. Then A is diagonalizable if and only if, for each eigenvalue a of A, geom-mult$(a) = $ alg-mult(a). In that case, $A = PJP^{-1}$ where J is a diagonal matrix whose entries are the eigenvalues of A, each appearing according to its algebraic multiplicity, and P is a matrix whose columns are eigenvectors forming bases for the associated eigenspaces.*

Proof. First suppose that for each eigenvalue a of A, geom-mult$(a) =$ alg-mult(a). In the notation of the proof of Theorem 1.14 in Chapter 1, $\mathcal{B} = \{v_1, \ldots, v_n\}$ is a basis of $\mathbb{C}^n$. Then, by Corollary 1.9, $A = P[\mathcal{T}_A]_\mathcal{B} P^{-1}$. But $\mathcal{B}$ is a basis of eigenvectors, so for each i, $\mathcal{T}_A(v_i) = Av_i = a_i v_i = 0v_1 + \ldots + 0v_{i-1} + a_i v_i + 0v_{i+1} + \ldots + 0v_n$. Then

$$[\mathcal{T}_A(v_i)]_\mathcal{B} = \begin{bmatrix} 0 \\ \vdots \\ a_i \\ \vdots \\ 0 \end{bmatrix}$$

with a_i in the ith position and 0 elsewhere. But $[\mathcal{T}_A(v_i)]_\mathcal{B}$ is the ith column of the matrix $[\mathcal{T}_A]_\mathcal{B}$, so we see that $J = [\mathcal{T}_A]_\mathcal{B}$ is a diagonal matrix.

Conversely, if $J = [\mathcal{T}_A]_\mathcal{B}$ is a diagonal matrix, then the same computation shows that $Av_i = a_i v_i$, so for each i, v_i is an eigenvector of A with associated eigenvalue a_i. $\qquad \square$

Theorem 1.11.   *Let A be a k-by-k matrix and suppose that $\mathbb{C}^k$ has a basis $\{v_1, \ldots, v_k\}$ consisting of a single chain of generalized eigenvectors of length k associated to an eigenvalue a. Then*

$$A = PJP^{-1}$$

where

$$J = \begin{bmatrix} a & 1 & & & & \\ & a & 1 & & & \\ & & a & 1 & & \\ & & & \ddots & \ddots & \\ & & & & a & 1 \\ & & & & & a \end{bmatrix}$$

is a matrix consisting of a single Jordan block and

$$P = \left[\, v_1 \,\middle|\, v_2 \,\middle|\, \cdots \,\middle|\, v_k \,\right]$$

is a matrix whose columns are generalized eigenvectors forming a chain.

Proof. Let $\mathcal{B} = \{v_1, \ldots, v_k\}$. Then, by Corollary 1.9, $A = P[\mathcal{T}_A]_\mathcal{B} P^{-1}$. Now the ith column of $[\mathcal{T}_A]_\mathcal{B}$ is $[\mathcal{T}_A(v_i)]_\mathcal{B} = [Av_i]_\mathcal{B}$. By the definition of a chain,

$$\begin{aligned} Av_i &= (A - aI + aI)v_i \\ &= (A - aI)v_i + aIv_i \\ &= v_{i-1} + av_i \text{ for } i > 1, \ = av_i \text{ for } i = 1, \end{aligned}$$

so for $i > 1$

$$[Av_i]_{\mathcal{B}} = \begin{bmatrix} 0 \\ \vdots \\ 1 \\ a \\ 0 \\ \vdots \end{bmatrix}$$

with 1 in the $(i-1)^{st}$ position, a in the ith position, and 0 elsewhere, and $[Av_1]_{\mathcal{B}}$ is similar, except that a is in the 1st position (there is no entry of 1), and every other entry is 0. Assembling these vectors, we see that the matrix $[\mathcal{T}_A]_{\mathcal{B}} = J$ has the form of a single k-by-k Jordan block with diagonal entries equal to a. $\qquad \square$

Lemma 1.12. *Let a be an eigenvalue of a matrix A. Then*

$$1 \leq \text{geom-mult}(a) \leq \text{alg-mult}(a) \, .$$

Proof. By the definition of an eigenvalue, there is at least one eigenvector v with eigenvalue a, and so E_a contains the nonzero vector v, and hence $\dim(E_a) \geq 1$.

Now suppose that a has geometric multiplicity k, and let $\{v_1, \ldots, v_k\}$ be a basis for the eigenspace E_a. Extend this basis to a basis $\mathcal{B} = \{v_1, \ldots, v_k, v_{k+1}, \ldots, v_n\}$ of $\mathbb{C}^n$. Let $B = [\mathcal{T}_A]_{\mathcal{B}}$. Then

$$B = \left[b_1 \, \middle| \, b_2 \, \middle| \, \cdots \, \middle| \, b_n \right]$$

with $b_i = [\mathcal{T}_A(v_i)]_{\mathcal{B}}$. But for i between 1 and k, $\mathcal{T}_A(v_i) = Av_i = av_i$, so

$$b_i = [av_i]_{\mathcal{B}} = \begin{bmatrix} 0 \\ \vdots \\ a \\ \vdots \\ 0 \end{bmatrix}$$

with a in the ith position and 0 elsewhere. (For $i > k$, we do not know what b_i is.)

Now we compute the characteristic polynomial of B, $\det(\lambda I - B)$. From our computation of B, we see that, for i between 1 and k, the ith column of $(\lambda I - B)$ is

$$\begin{bmatrix} 0 \\ \vdots \\ \lambda - a \\ \vdots \\ 0 \end{bmatrix}$$

with $\lambda - a$ in the ith position and 0 elsewhere. (For $i > k$, we do not know what the ith column of this matrix is.)

To compute the characteristic polynomial, i.e., the determinant of this matrix, we successively expand by minors of the $1^{\text{st}}, 2^{\text{nd}}, \ldots, k^{\text{th}}$ columns. Each of these gives a factor of $(\lambda - a)$, so we see that $\det(\lambda I - B) = (\lambda - a)^k q(\lambda)$ for some (unknown) polynomial $q(\lambda)$.

We have computed the characteristic polynomial of B, but what we need to know is the characteristic polynomial of A. But these are equal, as we see from the following computation (which uses the fact that scalar multiplication commutes with matrix multiplication, and properties of determinants):

$$\begin{aligned} \det(\lambda I - A) &= \det(\lambda I - PBP^{-1}) = \det(\lambda(PIP^{-1}) - PBP^{-1}) \\ &= \det(P(\lambda I)P^{-1} - PBP^{-1}) = \det(P(\lambda I - B)P^{-1}) \\ &= \det(P)\det(\lambda I - B)\det(P^{-1}) = \det(P)\det(\lambda I - B)(1/\det(P)) \\ &= \det(\lambda I - B). \end{aligned}$$

Thus, $\det(\lambda I - A)$, the characteristic polynomial of A, is divisible by $(\lambda - a)^k$ (and perhaps by a higher power of $(\lambda - a)$, and perhaps not, as we do not know anything about the polynomial $q(\lambda)$), so alg-mult$(a) \geq k =$ geom-mult(a), as claimed.   □

A.2   PROPERTIES OF THE COMPLEX EXPONENTIAL

In this section of the Appendix, we prove properties of the complex exponential. For convenience, we restate the basic definition and the properties we are trying to prove.

Definition 2.1.   For a complex number z, the exponential e^z is defined by

$$e^z = 1 + z + z^2/2! + z^3/3! + \ldots .$$

First we note that this definition indeed makes sense, as this power series converges for every complex number z. Now for the properties we wish to prove. Note that properties (2) and (3) are direct generalizations of the situation for the real exponential function.

Theorem 2.2.   *(1) (Euler) For any θ,*

$$e^{i\theta} = \cos(\theta) + i\sin(\theta).$$

(2) For any a,

$$\frac{d}{dz}(e^{az}) = ae^{az} .$$

(3) For any z_1 and z_2,

$$e^{z_1 + z_2} = e^{z_1}e^{z_2} .$$

(4) If $z = s + it$, then

$$e^z = e^s(\cos(t) + i\sin(t)) .$$

(5) For any z,

$$e^{\bar{z}} = \overline{e^z} .$$

Proof. (1) We begin with the following observation: $i^0 = 1$, $i^1 = i$, $i^2 = -1$, $i^3 = i^2i = -i$, $i^4 = i^3i = 1$, $i^5 = i^4i = i$, $i^6 = i^4i^2 = -1$, $i^7 = i^4i^3 = -i$, etc. In other words. the powers of i, beginning with i^0, successively cycle through $1, i, -1$, and $-i$. With this in mind, we compute directly from the definition:

$$e^{i\theta} = 1 + i\theta + (i\theta)^2/2! + (i\theta)^3/3! + (i\theta)^4/4! + (i\theta)^5/5! + (i\theta)^6/6! + (i\theta)^7/7! + \ldots$$
$$= 1 + i\theta - \theta^2/2! - i\theta^3/3! + \theta^4/4! + i\theta^5/5! - \theta^6/6! - i\theta^7/7! + \ldots .$$

We now rearrange the terms, gathering the terms that do not involve i together and gathering the terms that do involve i together. (That we can do so requires proof, but we shall not give that proof here.) We obtain:

$$e^{i\theta} = (1 - \theta^2/2! + \theta^4/4! - \theta^6/6! + \ldots) + i(\theta - \theta^3/3! + \theta^5/5! - \theta^7/7! + \ldots) .$$

But we recognize the power series inside the first set of parentheses as the power series for $\cos(\theta)$, and the power series inside the second set of parentheses as the power series for $\sin(\theta)$, completing the proof.

(2) We prove a more general formula. Consider the function e^{c+dz}, where c and d are arbitrary constants. To differentiate this function, we substitute in the power series and differentiate term-by-term (again a procedure that requires justification, but whose justification we again skip). Using the chain rule to take the derivative of each term, we obtain

$$e^{c+dz} = 1 + (c + dz) + (c + dz)^2/2! + (c + dz)^3/3! + (c + dz)^4/4! + \ldots$$
$$(e^{c+dz})' = d + 2d(c + dz)/2! + 3d(c + dz)^2/3! + 4d(c + dz)^3/4! + \ldots$$
$$= d + d(c + dz) + d(c + dz)^2/2! + d(c + dz)^3/3! + \ldots$$
$$= d(1 + (c + dz) + (c + dz)^2/2! + (c + dz)^3/3! + \ldots)$$
$$= de^{c+dz} .$$

Now, setting $c = 0$ and $d = a$, we obtain (2).

(3) Setting $c = a$ and $d = 1$ in the above formula, we find that $(e^{a+z})' = e^{a+z}$. In other words, $f_1(z) = e^{a+z}$ is a solution of the differential equation $f'(z) = f(z)$, and $f_1(0) = e^{a+0} = e^a$. On the other hand, setting $f_2(z) = e^a e^z$, we see from (2), and the fact that e^a is a constant, that $f_2(z)$ is also a solution of the differential equation $f'(z) = f(z)$, and $f_2(0) = e^a e^0 = e^a$. Thus, $f_1(z)$ and $f_2(z)$ are solutions of the same first-order linear differential equation satisfying the same initial condition, so by the fundamental existence and uniqueness theorem (also valid for complex functions), they must be equal. Thus, $e^{a+z} = e^a e^z$. Setting $a = z_1$ and $z = z_2$, we obtain (3).

(4) This follows directly from (1) and (3):

$$e^z = e^{s+it} = e^s e^{it} = e^s (\cos(t) + i \sin(t)) .$$

(5) This follows directly from (1) and (4). Let $z = s + it$, so $\bar{z} = s - it$. We compute:

$$e^{\bar{z}} = e^{s-it} = e^{s+i(-t)} = e^s (\cos(-t) + i \sin(-t)) = e^s (\cos(t) + i(-\sin(t)))$$
$$= e^s (\cos(t) - i \sin(t)) = \overline{e^s (\cos(t) + i \sin(t))} = \overline{e^z} .$$

$\square$

APPENDIX B

Answers to Odd-Numbered Exercises

Chapter 1

1. $A = \begin{bmatrix} -7 & 4 \\ 9 & -5 \end{bmatrix} \begin{bmatrix} 3 & 0 \\ 0 & 5 \end{bmatrix} \begin{bmatrix} -7 & 4 \\ 9 & -5 \end{bmatrix}^{-1}$.

3. $A = \begin{bmatrix} -21 & 1 \\ -49 & 0 \end{bmatrix} \begin{bmatrix} 3 & 1 \\ 0 & 3 \end{bmatrix} \begin{bmatrix} -21 & 1 \\ -49 & 0 \end{bmatrix}^{-1}$.

5. $A = \begin{bmatrix} -5 & 1 \\ -25 & 0 \end{bmatrix} \begin{bmatrix} 7 & 1 \\ 0 & 7 \end{bmatrix} \begin{bmatrix} -5 & 1 \\ -25 & 0 \end{bmatrix}^{-1}$.

7. $A = \begin{bmatrix} 0 & 2 & 0 \\ 1 & 1 & -3 \\ 1 & 1 & 1 \end{bmatrix} \begin{bmatrix} -1 & 0 & 0 \\ 0 & 1 & 0 \\ 0 & 0 & 3 \end{bmatrix} \begin{bmatrix} 0 & 2 & 0 \\ 1 & 1 & -3 \\ 1 & 1 & 1 \end{bmatrix}^{-1}$.

9. $A = \begin{bmatrix} -1 & -2 & -2 \\ 1 & 0 & -1 \\ 0 & 1 & 1 \end{bmatrix} \begin{bmatrix} -3 & 0 & 0 \\ 0 & -3 & 0 \\ 0 & 0 & 1 \end{bmatrix} \begin{bmatrix} -1 & -2 & -2 \\ 1 & 0 & -1 \\ 0 & 1 & 1 \end{bmatrix}^{-1}$.

11. $A = \begin{bmatrix} -1 & -2 & -1 \\ 0 & 1 & 1 \\ 1 & 0 & 1 \end{bmatrix} \begin{bmatrix} 4 & 0 & 0 \\ 0 & 4 & 0 \\ 0 & 0 & 2 \end{bmatrix} \begin{bmatrix} -1 & -2 & -1 \\ 0 & 1 & 1 \\ 1 & 0 & 1 \end{bmatrix}^{-1}$.

13. $A = \begin{bmatrix} -1 & 0 & 0 \\ -1 & 0 & 1 \\ 0 & 1 & 1 \end{bmatrix} \begin{bmatrix} -2 & 1 & 0 \\ 0 & -2 & 0 \\ 0 & 0 & 4 \end{bmatrix} \begin{bmatrix} -1 & 0 & 0 \\ -1 & 0 & 1 \\ 0 & 1 & 1 \end{bmatrix}^{-1}$.

15. $A = \begin{bmatrix} -3 & -1 & -2 \\ -2 & -1 & -2 \\ 3 & 1 & 3 \end{bmatrix} \begin{bmatrix} 0 & 1 & 0 \\ 0 & 0 & 0 \\ 0 & 0 & 3 \end{bmatrix} \begin{bmatrix} -3 & -1 & -2 \\ -2 & -1 & -2 \\ 3 & 1 & 3 \end{bmatrix}^{-1}$.

17. $A = \begin{bmatrix} -2 & 1 & 1 \\ -10 & 0 & 2 \\ -6 & 0 & 0 \end{bmatrix} \begin{bmatrix} 1 & 1 & 0 \\ 0 & 1 & 0 \\ 0 & 0 & 1 \end{bmatrix} \begin{bmatrix} -2 & 1 & 1 \\ -10 & 0 & 2 \\ -6 & 0 & 0 \end{bmatrix}^{-1}$.

19. $A = \begin{bmatrix} 1 & 2 & 0 \\ 2 & 3 & 0 \\ 1 & 3 & 1 \end{bmatrix} \begin{bmatrix} 0 & 1 & 0 \\ 0 & 0 & 1 \\ 0 & 0 & 0 \end{bmatrix} \begin{bmatrix} 1 & 2 & 0 \\ 2 & 3 & 0 \\ 1 & 3 & 1 \end{bmatrix}^{-1}$.

Section 2.1

1a. $Y = \begin{bmatrix} -7c_1 e^{3x} + 4c_2 e^{5x} \\ 9c_1 e^{3x} - 5c_2 e^{5x} \end{bmatrix}$.   b. $Y = \begin{bmatrix} -7e^{3x} + 8e^{5x} \\ 9e^{3x} - 10e^{5x} \end{bmatrix}$.

3a. $Y = \begin{bmatrix} (-21c_1 + c_2)e^{3x} - 21c_2 x e^{3x} \\ -49c_1 e^{3x} \quad\quad - 49c_2 x e^{3x} \end{bmatrix}$.   b. $Y = \begin{bmatrix} 41e^{3x} + 21x e^{3x} \\ 98e^{3x} + 49x e^{3x} \end{bmatrix}$.

5a. $Y = \begin{bmatrix} (-5c_1 + c_2)e^{7x} - 5c_2 x e^{7x} \\ -25c_1 e^{7x} \quad\quad - 25c_2 x e^{7x} \end{bmatrix}$.   b. $Y = \begin{bmatrix} -10e^{7x} - 25x e^{7x} \\ -75e^{7x} - 125x e^{7x} \end{bmatrix}$.

7a. $Y = \begin{bmatrix} 2c_2 e^x \\ c_1 e^{-x} + c_2 e^x - 3c_3 e^{3x} \\ c_1 e^{-x} + c_2 e^x + c_3 e^{3x} \end{bmatrix}$.   b. $Y = \begin{bmatrix} 6e^x \\ 2e^{-x} + 3e^x - 15e^{3x} \\ 2e^{-x} + 3e^x + 5e^{3x} \end{bmatrix}$.

9a. $Y = \begin{bmatrix} (-c_1 - 2c_2)e^{-3x} - 2c_3 e^x \\ c_1 e^{-3x} - c_3 e^x \\ c_2 e^{-3x} + c_3 e^x \end{bmatrix}$.   b. $Y = \begin{bmatrix} 0 \\ 2e^{-3x} \\ -e^{-3x} \end{bmatrix}$.

11a. $Y = \begin{bmatrix} (-c_1 - 2c_2)e^{4x} - c_3 e^{2x} \\ c_2 e^{4x} + c_3 e^{2x} \\ c_1 e^{4x} + c_3 e^{2x} \end{bmatrix}$.   b. $Y = \begin{bmatrix} 2e^{4x} - 5e^{2x} \\ -3e^{4x} + 5e^{2x} \\ 4e^{4x} + 5e^{2x} \end{bmatrix}$.

13a. $Y = \begin{bmatrix} -c_1 e^{-2x} - c_2 x e^{-2x} \\ -c_1 e^{-2x} - c_2 x e^{-2x} + c_3 e^{4x} \\ c_2 e^{-2x} \quad\quad + c_3 e^{4x} \end{bmatrix}$.   b. $Y = \begin{bmatrix} -e^{-2x} - 2x e^{-2x} \\ -e^{-2x} - 2x e^{-2x} + 4e^{4x} \\ 2e^{-2x} \quad\quad + 4e^{4x} \end{bmatrix}$.

15a. $Y = \begin{bmatrix} (-3c_1 - c_2) - 3c_2 x - 2c_3 e^{3x} \\ (-2c_1 - c_2) - 2c_2 x - 2c_3 e^{3x} \\ (3c_1 + c_2) + 3c_2 x + 3c_3 e^{3x} \end{bmatrix}$.   b. $Y = \begin{bmatrix} -9 - 9x + 10e^{3x} \\ -7 - 6x + 10e^{3x} \\ 9 + 9x - 15e^{3x} \end{bmatrix}$.

17a. $Y = \begin{bmatrix} (-2c_1 + c_2 + c_3)e^x - 2c_2xe^x \\ (-10c_1 + 2c_3)e^x - 10c_2xe^x \\ -6c_1e^x - 6c_2xe^x \end{bmatrix}$.   b. $Y = \begin{bmatrix} 3e^x - 14xe^x \\ 10e^x - 70xe^x \\ 18e^x - 42xe^x \end{bmatrix}$.

19a. $Y = \begin{bmatrix} (c_1 + 2c_2) & + (c_2 + 2c_3)x + (c_3/2)x^2 \\ (2c_1 + 3c_2) & + (2c_2 + 3c_3)x + c_3x^2 \\ (c_1 + 3c_2 + c_3) + (c_2 + 3c_3)x + (c_3/2)x^2 \end{bmatrix}$.   b. $Y = \begin{bmatrix} 6 + 5x + x^2 \\ 11 + 8x + 2x^2 \\ 9 + 7x + x^2 \end{bmatrix}$.

Section 2.2

1a.

$$Y = \begin{bmatrix} e^{4x}(\cos(3x) + 3\sin(3x)) & e^{4x}(-3\cos(3x) + \sin(3x)) \\ 2e^{4x}\cos(3x) & 2e^{4x}\sin(3x) \end{bmatrix} \begin{bmatrix} c_1 \\ c_2 \end{bmatrix}$$
$$= \begin{bmatrix} (c_1 - 3c_2)e^{4x}\cos(3x) + (3c_1 + c_2)e^{4x}\sin(3x) \\ 2c_1e^{4x}\cos(3x) + 2c_2e^{4x}\sin(3x) \end{bmatrix}.$$

b. $Y = \begin{bmatrix} 8e^{4x}\cos(3x) + 19e^{4x}\sin(3x) \\ 13e^{4x}\cos(3x) - \sin(3x) \end{bmatrix}$.

3a.

$$Y = \begin{bmatrix} e^{7x}(2\cos(3x) + 3\sin(3x)) & e^{7x}(-3\cos(3x) + 2\sin(3x)) \\ e^{7x}\cos(3x) & e^{7x}\sin(3x) \end{bmatrix} \begin{bmatrix} c_1 \\ c_2 \end{bmatrix}$$
$$= \begin{bmatrix} (2c_1 - 3c_2)e^{7x}\cos(3x) + (3c_1 + 2c_2)e^{7x}\sin(3x) \\ c_1e^{7x}\cos(3x) + c_2e^{7x}\sin(3x) \end{bmatrix}.$$

b. $Y = \begin{bmatrix} 2e^{7x}\cos(3x) + 3e^{7x}\sin(3x) \\ e^{7x}\cos(3x) \end{bmatrix}$.

5.

$$Y = \begin{bmatrix} e^{2x}(-\cos(5x) - \sin(5x)) & e^{2x}(\cos(5x) - \sin(5x)) & 0 \\ e^{2x}(-3\cos(5x) + 4\sin(5x)) & e^{2x}(-4\cos(5x) - 3\sin(5x)) & -2e^{3x} \\ 3e^{2x}\cos(5x) & 3e^{2x}\sin(5x) & e^{3x} \end{bmatrix} \begin{bmatrix} c_1 \\ c_2 \\ c_3 \end{bmatrix}$$
$$= \begin{bmatrix} (-c_1 + c_2)e^{2x}\cos(5x) + (-c_1 - c_2)e^{2x}\sin(5x) \\ (-3c_1 - 4c_2)e^{2x}\cos(5x) + (4c_1 - 3c_2)e^{2x}\sin(5x) - 2c_3e^{3x} \\ 3c_1e^{2x}\cos(5x) + 3c_2e^{2x}\sin(5x) + c_3e^{3x} \end{bmatrix}.$$

Section 2.3

1. $Y_i = \begin{bmatrix} 10e^{8x} - 168e^{4x} \\ -12e^{8x} + 213e^{4x} \end{bmatrix}$.

3. $Y_i = \begin{bmatrix} -20e^{4x} + 9e^{5x} \\ -49e^{4x} + 23e^{5x} \end{bmatrix}.$

5. $Y_i = \begin{bmatrix} -2e^{10x} + e^{12x} \\ -25e^{10x} + 10e^{12x} \end{bmatrix}.$

7. $Y_i = \begin{bmatrix} -1 \\ -1 - 2e^{2x} - 3e^{4x} \\ -1 + e^{2x} + 2e^{4x} \end{bmatrix}.$

Section 2.4

1a. $e^{Ax} = \begin{bmatrix} -35e^{3x} + 36e^{5x} & -28e^{3x} + 28e^{5x} \\ 45e^{3x} - 45e^{5x} & 36e^{3x} - 35e^{5x} \end{bmatrix}.$

$Y = \begin{bmatrix} (-35\gamma_1 - 28\gamma_2)e^{3x} + (36\gamma_1 + 28\gamma_2)e^{5x} \\ (45\gamma_1 + 36\gamma_2)e^{3x} + (-45\gamma_1 - 35\gamma_2)e^{5x} \end{bmatrix}.$

3a. $e^{Ax} = \begin{bmatrix} e^{3x} - 21xe^{3x} & 9xe^{3x} \\ -49xe^{3x} & e^{3x} + 21xe^{3x} \end{bmatrix}.$

$Y = \begin{bmatrix} \gamma_1 e^{3x} + (-21\gamma_1 + 9\gamma_2)xe^{3x} \\ \gamma_2 e^{3x} + (-49\gamma_1 + 21\gamma_2)xe^{3x} \end{bmatrix}.$

5a. $e^{Ax} = \begin{bmatrix} e^{7x} - 5xe^{7x} & xe^{7x} \\ -25xe^{7x} & e^{7x} + 5xe^{7x} \end{bmatrix}.$

$Y = \begin{bmatrix} \gamma_1 e^{7x} + (-5\gamma_1 + \gamma_2)xe^{7x} \\ \gamma_2 e^{7x} + (-25\gamma_1 + 5\gamma_2)xe^{7x} \end{bmatrix}.$

7a. $e^{Ax} = \begin{bmatrix} e^x & 0 & 0 \\ -\frac{1}{2}e^{-x} + \frac{1}{2}e^x & \frac{1}{4}e^{-x} + \frac{3}{4}e^{3x} & \frac{3}{4}e^{-x} - \frac{3}{4}e^{3x} \\ -\frac{1}{2}e^{-x} + \frac{1}{2}e^x & \frac{1}{4}e^{-x} - \frac{1}{4}e^{3x} & \frac{3}{4}e^{-x} + \frac{1}{4}e^{3x} \end{bmatrix}.$

$Y = \begin{bmatrix} \gamma_1 e^x \\ (-\frac{1}{2}\gamma_1 + \frac{1}{4}\gamma_2 + \frac{3}{4}\gamma_3)e^{-x} + \frac{1}{2}\gamma_1 e^x + (\frac{3}{4}\gamma_2 - \frac{3}{4}\gamma_3)e^{3x} \\ (-\frac{1}{2}\gamma_1 + \frac{1}{4}\gamma_2 + \frac{3}{4}\gamma_3)e^{-x} + \frac{1}{2}\gamma_1 e^x + (-\frac{1}{4}\gamma_2 + \frac{1}{4}\gamma_3)e^{3x} \end{bmatrix}.$

9a. $\quad e^{Ax} = \begin{bmatrix} -e^{-3x} + 2e^x & -2e^{-3x} + 2e^x & -4e^{-3x} + 4e^x \\ -e^{-3x} + e^x & e^x & -2e^{-3x} + 2e^x \\ e^{-3x} - e^x & e^{-3x} - e^x & 3e^{-3x} - 2e^x \end{bmatrix}.$

$$Y = \begin{bmatrix} (-\gamma_1 - 2\gamma_2 - 4\gamma_3)e^{-3x} + (2\gamma_1 + 2\gamma_2 + 4\gamma_3)e^x \\ (-\gamma_1 - 2\gamma_3)e^{-3x} + (\gamma_1 + \gamma_2 + 2\gamma_3)e^x \\ (\gamma_1 + \gamma_2 + 2\gamma_3)e^{-3x} + (-\gamma_1 - \gamma_2 - 2\gamma_3)e^x \end{bmatrix}.$$

11a. $\quad e^{Ax} = \begin{bmatrix} \frac{3}{2}e^{4x} - \frac{1}{2}e^{2x} & e^{4x} - e^{2x} & \frac{1}{2}e^{4x} - \frac{1}{2}e^{2x} \\ -\frac{1}{2}e^{4x} + \frac{1}{2}e^{2x} & e^{2x} & -\frac{1}{2}e^{4x} + \frac{1}{2}e^{2x} \\ -\frac{1}{2}e^{4x} + \frac{1}{2}e^{2x} & -e^{4x} + e^{2x} & \frac{1}{2}e^{4x} + \frac{1}{2}e^{2x} \end{bmatrix}.$

$$Y = \begin{bmatrix} (\frac{3}{2}\gamma_1 + \gamma_2 + \frac{1}{2}\gamma_3)e^{4x} + (-\frac{1}{2}\gamma_1 - \gamma_2 - \frac{1}{2}\gamma_3)e^{2x} \\ (-\frac{1}{2}\gamma_1 - \frac{1}{2}\gamma_3)e^{4x} + (\frac{1}{2}\gamma_1 + \gamma_2 + \frac{1}{2}\gamma_3)e^{2x} \\ (-\frac{1}{2}\gamma_1 - \gamma_2 + \frac{1}{2}\gamma_3)e^{4x} + (\frac{1}{2}\gamma_1 + \gamma_2 + \frac{1}{2}\gamma_3)e^{2x} \end{bmatrix}.$$

13a. $\quad e^{Ax} = \begin{bmatrix} e^{-2x} - xe^{-2x} & xe^{-2x} & -xe^{-2x} \\ e^{-2x} - xe^{-2x} - e^{4x} & xe^{-2x} + e^{4x} & -xe^{-2x} \\ e^{-2x} - e^{4x} & -e^{-2x} + e^{4x} & e^{-2x} \end{bmatrix}.$

$$Y = \begin{bmatrix} \gamma_1 e^{-2x} + (-\gamma_1 + \gamma_2 - \gamma_3)xe^{-2x} \\ \gamma_1 e^{-2x} + (-\gamma_1 + \gamma_2 - \gamma_3)xe^{-2x} + (-\gamma_1 + \gamma_2)e^{4x} \\ (\gamma_1 - \gamma_2 + \gamma_3)e^{-2x} + (-\gamma_1 + \gamma_2)e^{4x} \end{bmatrix}.$$

15a. $\quad e^{Ax} = \begin{bmatrix} 3 - 2e^{3x} & 9x & 2 + 6x - 2e^{3x} \\ 2 - 2e^{3x} & 1 + 6x & 2 + 4x - 2e^{3x} \\ -3 + 3e^{3x} & -9x & -2 - 6x + 3e^{3x} \end{bmatrix}.$

$$Y = \begin{bmatrix} (3\gamma_1 + 2\gamma_3) + (9\gamma_2 + 6\gamma_3)x + (-2\gamma_1 - 2\gamma_3)e^{3x} \\ (2\gamma_1 + \gamma_2 + 2\gamma_3) + (6\gamma_2 + 4\gamma_3)x + (-2\gamma_1 - 2\gamma_3)e^{3x} \\ (-3\gamma_1 - 2\gamma_3) + (-9\gamma_2 - 6\gamma_3)x + (3\gamma_1 + 3\gamma_3)e^{3x} \end{bmatrix}.$$

17a. $\quad e^{Ax} = \begin{bmatrix} e^x - 2xe^x & xe^x & -xe^x \\ -10xe^x & e^x + 5xe^x & -5xe^x \\ -6xe^x & 3xe^x & e^x - 3xe^x \end{bmatrix}.$

$$Y = \begin{bmatrix} \gamma_1 e^x + (-2\gamma_1 + \gamma_2 - \gamma_3)xe^x \\ \gamma_2 e^x + (-10\gamma_1 + 5\gamma_2 - 5\gamma_3)xe^x \\ \gamma_3 e^x + (-6\gamma_1 + 3\gamma_2 - 3\gamma_3)xe^x \end{bmatrix}.$$

19a. $e^{Ax} = \begin{bmatrix} 1 - 4x - \frac{3}{2}x^2 & x + \frac{1}{2}x^2 & 2x + \frac{1}{2}x^2 \\ -5x - 3x^2 & 1 + x + x^2 & 3x + x^2 \\ -7x - \frac{3}{2}x^2 & 2x + \frac{1}{2}x^2 & 1 + 3x + \frac{1}{2}x^2 \end{bmatrix}.$

$$Y = \begin{bmatrix} \gamma_1 + (-4\gamma_1 + \gamma_2 + 2\gamma_3)x + (-\frac{3}{2}\gamma_1 + \frac{1}{2}\gamma_2 + \frac{1}{2}\gamma_3)x^2 \\ \gamma_2 + (-5\gamma_1 + \gamma_2 + 3\gamma_3)x + (-3\gamma_1 + \gamma_2 + \gamma_3)x^2 \\ \gamma_3 + (-7\gamma_1 + 2\gamma_2 + 3\gamma_3)x + (-\frac{3}{2}\gamma_1 + \frac{1}{2}\gamma_2 + \frac{1}{2}\gamma_3)x^2 \end{bmatrix}.$$

21a. $e^{Ax} = \begin{bmatrix} e^{4x}(\cos(3x) - \frac{1}{3}\sin(3x)) & \frac{5}{3}e^{4x}\sin(3x) \\ -\frac{2}{3}e^{4x}\sin(3x) & e^{4x}(\cos(3x) + \frac{1}{3}\sin(3x)) \end{bmatrix}.$

$$Y = \begin{bmatrix} \gamma_1 e^{4x}\cos(3x) + (-\frac{1}{3}\gamma_1 + \frac{5}{3}\gamma_2)e^{4x}\sin(3x) \\ \gamma_2 e^{4x}\cos(3x) + (-\frac{2}{3}\gamma_1 + \frac{1}{3}\gamma_2)e^{4x}\sin(3x) \end{bmatrix}.$$

23a. $e^{Ax} = \begin{bmatrix} e^{7x}(\cos(3x) - \frac{2}{3}\sin(3x)) & \frac{13}{3}e^{7x}\sin(3x) \\ -\frac{1}{3}e^{7x}\sin(3x) & e^{7x}(\cos(3x) + \frac{2}{3}\sin(3x)) \end{bmatrix}.$

$$Y = \begin{bmatrix} \gamma_1 e^{7x}\cos(3x) + (-\frac{2}{3}\gamma_1 + \frac{13}{3}\gamma_2)e^{7x}\sin(3x) \\ \gamma_2 e^{7x}\cos(3x) + (-\frac{1}{3}\gamma_1 + \frac{2}{3}\gamma_2)e^{7x}\sin(3x) \end{bmatrix}.$$

Index

Printed in the United States
by Baker & Taylor Publisher Services